KENT'S SEASIDE RESORTS

THROUGH TIME

John Clancy

AMBERLEY PUBLISHING

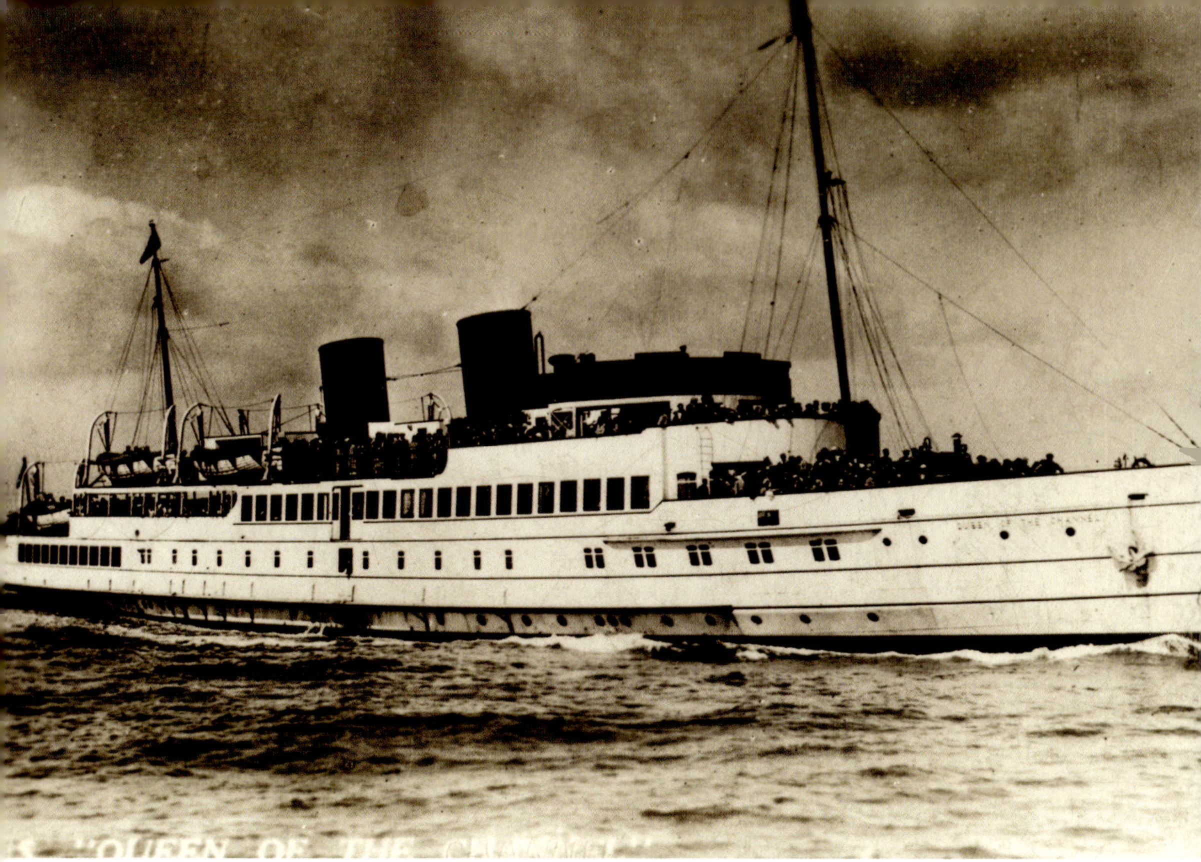

Queen of the Channel

The New Medway Steam Packet *Queen of the Channel* seen here fully laden. By 1937 this was the kind of vessel used by many day-trippers to take them to the coast.

First published 2012

Amberley Publishing
The Hill, Stroud, Gloucestershire, GL5 4EP
www.amberley-books.com

Copyright © John Clancy, 2012

The right of John Clancy to be identified as the Author of this work has been asserted in accordance with the Copyrights, Designs and Patents Act 1988.

ISBN 978 1 4456 0363 6

British Library Cataloguing in Publication Data.
A catalogue record for this book is available from the British Library.

Typesetting by Amberley Publishing.
Printed in Great Britain.

Introduction

How best can you define a seaside resort? For the most part, they are towns that developed from small fishing communities at the height of the late-eighteenth-century fashion for sea-bathing. Kent was at the forefront in the growth of the British seaside industry from the eighteenth century onwards, largely due to its extensive coastline, one of the longest in Britain. Being within relatively easy access of London, a trip to the coast became a great attraction for fashionable Londoners. Today we have the choice of travelling by car, train or coach, but how did the Georgians make the journey? Prior to the invention of the railway and a decent road network, those travelling to the coast went by hoys and later, steamships. Hoys were flat-bottomed sailing vessels that were ideally suited to working in shallow waters but took anything up to seventy-two hours to reach Margate from London. The first steamship service between London and Margate was introduced in 1815 and it made trade with London much faster. They could make at least nine voyages in the time a hoy could manage just one. By 1841 there were six different companies competing for the Margate passenger traffic of steamship excursions; the last service ran in 1967. But as some resorts did not have a harbour, there was often a problem with getting passengers ashore safely; a leaky rowing boat or the back of a sturdy fisherman usually had to suffice. Clearly landing stages were needed to disembark passengers safely and comfortably, and so the idea of a pier was conceived.

By the mid-nineteenth century the pier had become the most fashionable promenade in town. Its attraction was likened to being on a large ship, far out to sea, with the sea all around you and sea breezes blowing from all sides. From being simply a landing stage, piers had evolved into attractions in their own right, with ornate pavilions containing all sorts of amusements as well as tea rooms; later, concert halls were added to the larger piers. They quickly became an iconic emblem of the seaside experience, a unique resort attraction, and from the 1860s onward the new seaside resorts fiercely contested to have the best show-piece pleasure pier.

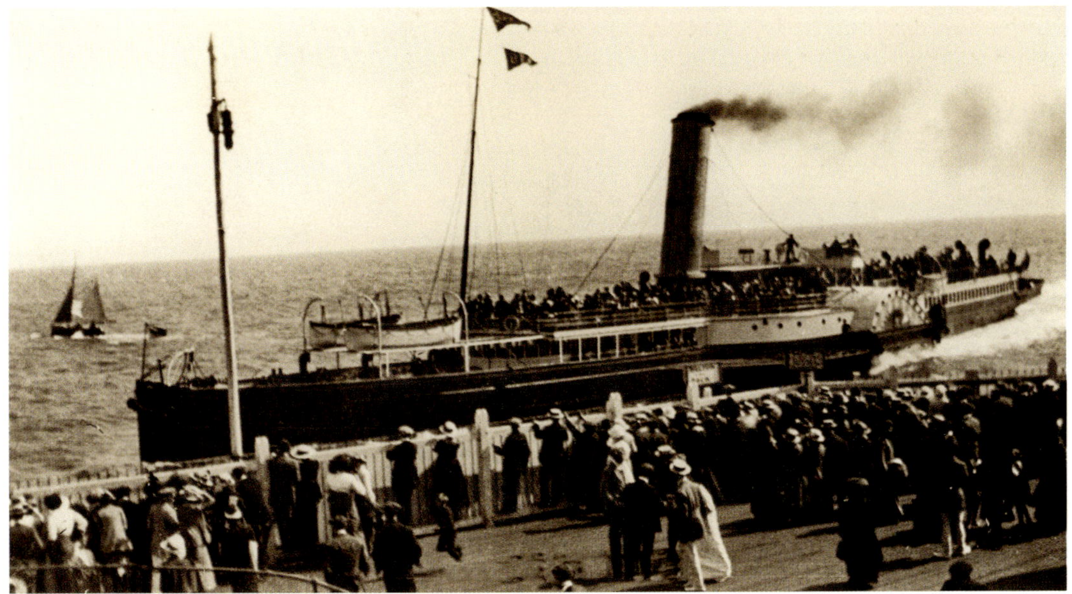

Day Tripper Special
Another 'day-tripper special' was the PS *Clacton Belle*, a paddle steamer on the Tilbury–Clacton–Margate–Southend–Rochester run, seen here in the 1920s arriving at the pier-head.

The pier-building era came to an end in around 1910, but by then almost 100 pleasure piers had been built all along the coastline. But piers cannot be taken for granted; for almost every pier built between 1860 and 1910, there was another that never got past the planning stage, such was the demand to build them. Almost every resort at one time or another had plans to build a pier yet many, especially the smaller resorts, were thwarted by a lack of finances, local opposition or indifference. It was never quite as simple as the proposal being placed before the local council planning committee. Consent had to be obtained from Parliament through a Parliamentary Bill and from the foreshore authority, usually the Board of Trade. The local lord of the manor might also be involved, as might other local bodies. And if all these hurdles were overcome and the finance was in place, there then followed the actual construction work in an environment that was fraught with difficulties and dangers.

These Victorian structures have stood intact in many cases for over a century or more, but now, through neglect, the weather and the general ravages of time, they are starting to show their age. In some cases the original pier has fallen down or been demolished for safety reasons, while others have been rebuilt or radically altered. In some resorts, the pier or jetty has not only vanished from the landscape but also from people's memory as well. All we are left with are subtle reminders in the naming of roads and avenues that once led to them, and it is this that makes books like this all the more relevant. This book includes a representative number of old and new photos of the many resorts there are in Kent, showing how each grew and developed over time. It is neither a comprehensive nor exhaustive collection of pictures; space precludes that. It will be seen that there were key features that had to be included to make the resort successful, like the pier, a clock tower, a bandstand and ornamental gardens.

This book will show how the character of the typical British seaside resort has changed in the past century and more. From the early nineteenth century until the 1960s, almost everyone enjoyed their annual seaside holiday at one of the many resorts throughout Britain, but then tastes changed as foreign holidays became more fashionable. Like many families in the 1940s and 1950s, I too enjoyed an annual holiday in Margate, and many of these early views bring back fond childhood memories. But looking at the modern-day equivalent views it shows how neglected this part of our heritage has now become.

The picture below shows another popular paddle steamer with day-trippers, the *Crested Eagle*, which sailed from London–Clacton–Southend–Margate. She was called upon to help rescue troops from the beaches at Dunkirk in 1940 where, sadly, she was sunk by enemy fire.

The inset shows a paddle steamer approaching Margate Pier, or Margate Jetty as it was then known. It allowed steamers to land their passengers at any state of the tide, and survived ship collisions, storms and several attempts at demolition before finally being pulled down in 1978.

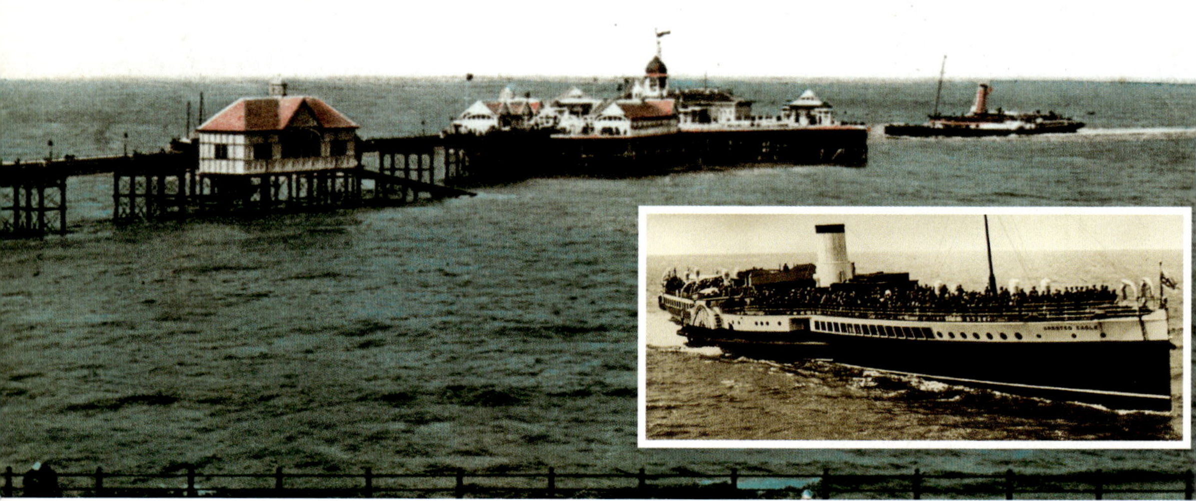

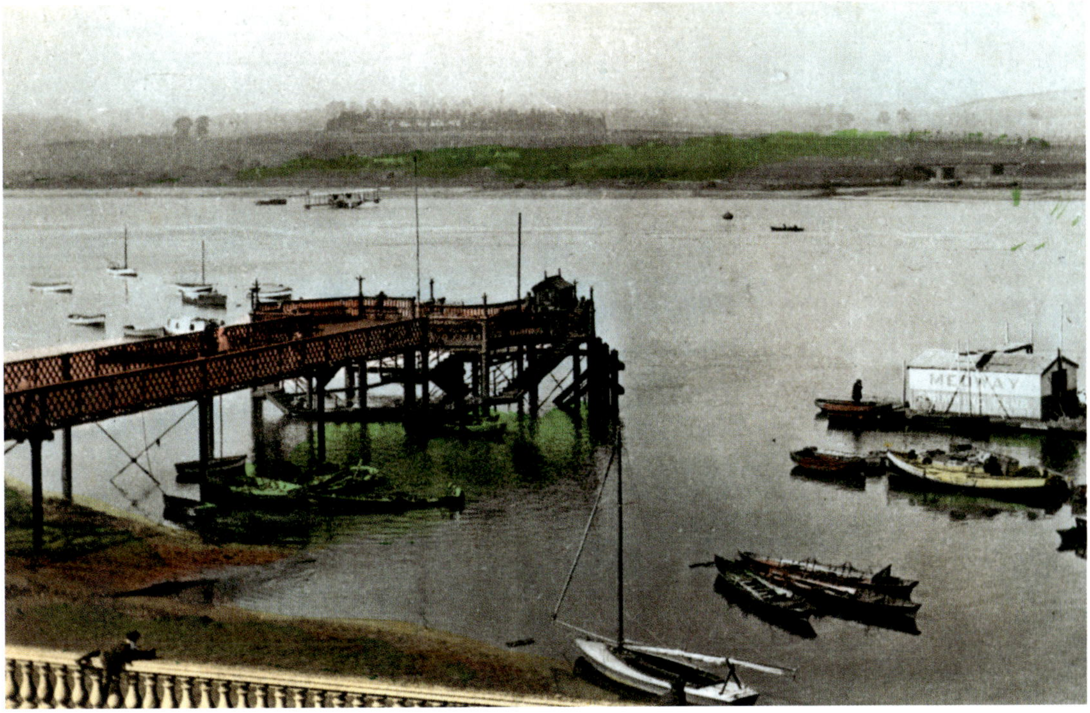

Rochester Pier

Rochester Pier was used principally for river traffic, but also used as an embarkation/disembarkation point for vessels taking day-trippers to the coast.

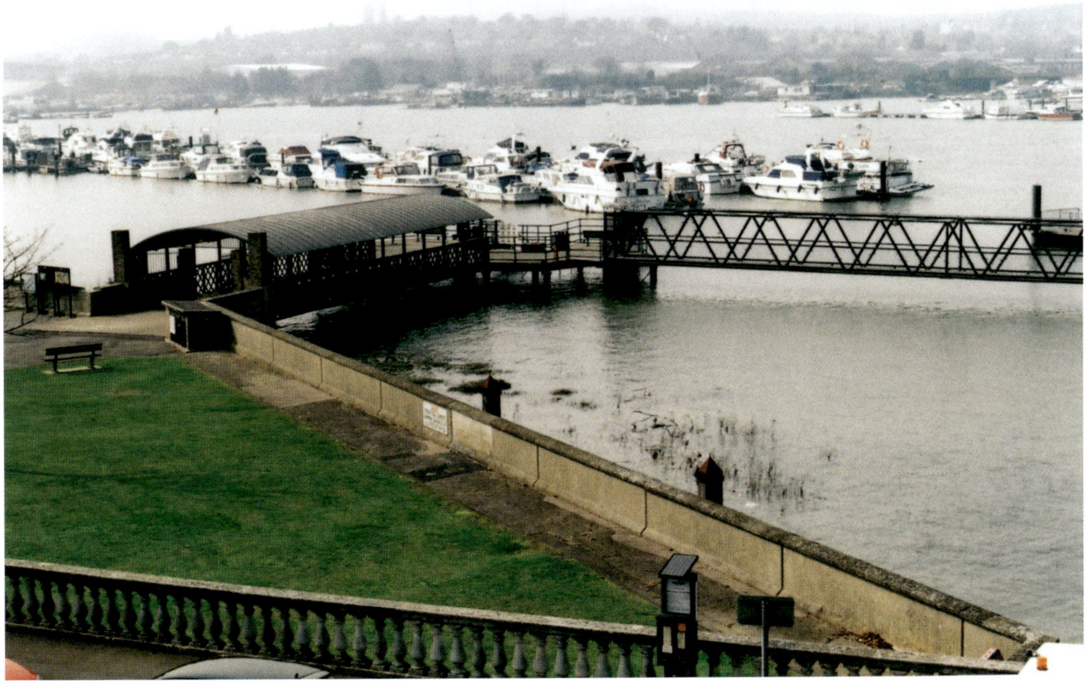

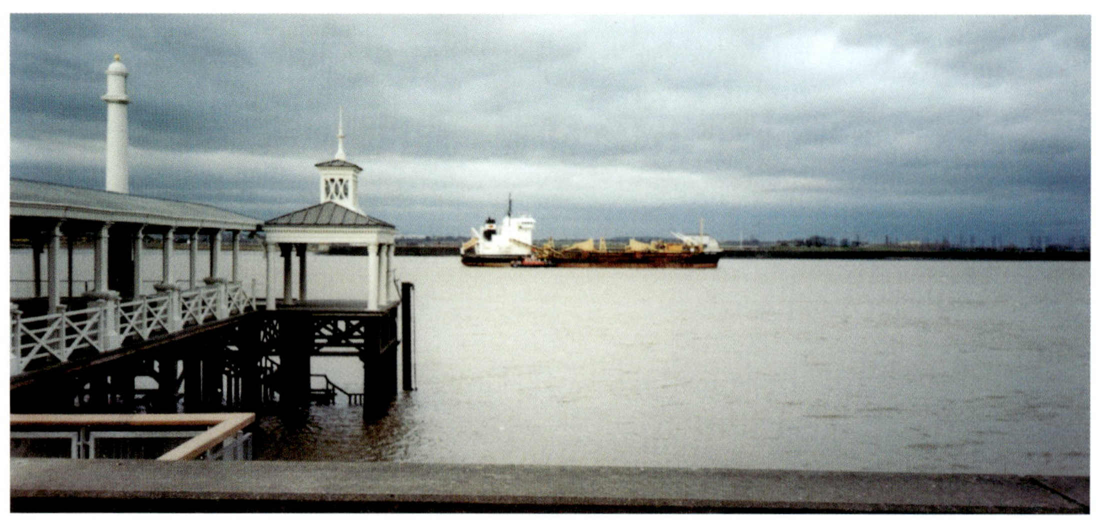

Gravesend & Rochester

Gravesend also had a pier for much the same reason as Rochester, but with a subtle difference – it also had a beach. In order to record as many of the county's piers and jetties, included in this book are those at places like Gravesend and Rochester, whose piers were never designed for recreational purposes. Piled pier structures like these were purely landing stages for river traffic, nothing more and nothing less. They were used, however, for embarking/disembarking passengers on and off ships that would take them to Southend, Clacton, Margate, etc. Equally the jetties at Whitstable and Broadstairs primarily served the local fishing fleet, as did the harbours at Ramsgate and Margate, but even this had a certain appeal to the new day-trippers and holidaymakers, as their souvenir postcards which they sent home to friends and family attest. Likewise, those at Folkestone and Dover primarily served those travelling across the Channel. The observant will notice that Sheerness-on-Sea has been omitted from this book. There are two interrelated reasons for this. Firstly, a lack of space, and secondly, Sheerness has been more than adequately covered in detail in my earlier book, *Isle of Sheppey Through Time*.

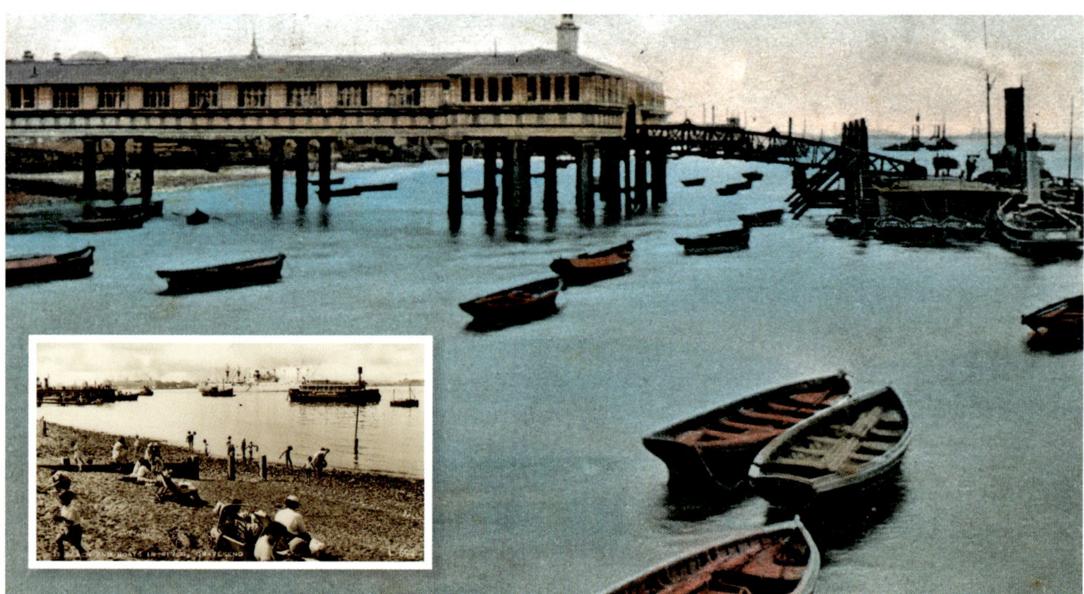

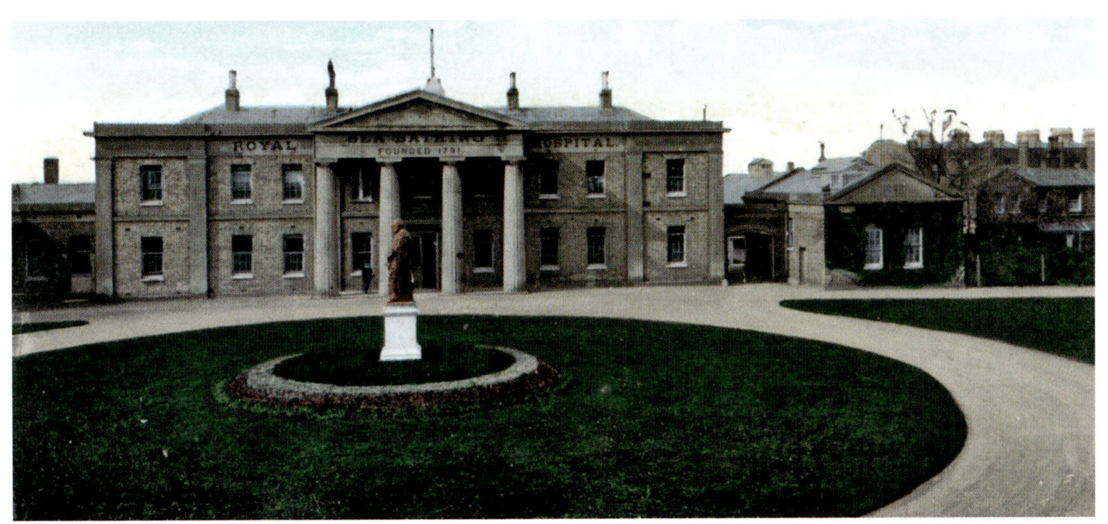

The Beginnings – The Royal Sea Bathing Hospital, Margate

Such was the Georgians' belief that sea bathing was beneficial to their health, a hospital was built at Margate in 1796 by Dr John Coakley Letttsom, a Quaker physician who is better remembered for being the founder of the Medical Society of London. His main concern was to find a cure for scrofula, or tuberculosis, by means of thalasso therapy, or the sea water cure. This treatment had been pioneered in Brighton by Dr Richard Russell on the fashionably rich, but Lettsom wanted to extend it to the deprived classes from London's East End, whose lifestyle made them susceptible to all forms of tuberculosis. Margate's fresh air and good food offered the perfect environment. Among the more well-known patients to have been admitted to this hospital was Karl Marx, who came here in 1866 to convalesce from an attack of boils. The hospital continued to treat surgical TB cases until the early 1950s, when improvements in treatment, preventative medicine and a rise in the standard of living made TB an uncommon disease, and the hospital started accepting general orthopaedic cases. As a hospital it closed in 1996 and the site was sold for redevelopment. It was decided to convert the old hospital into a complex of flats and work on this commenced in 2006.

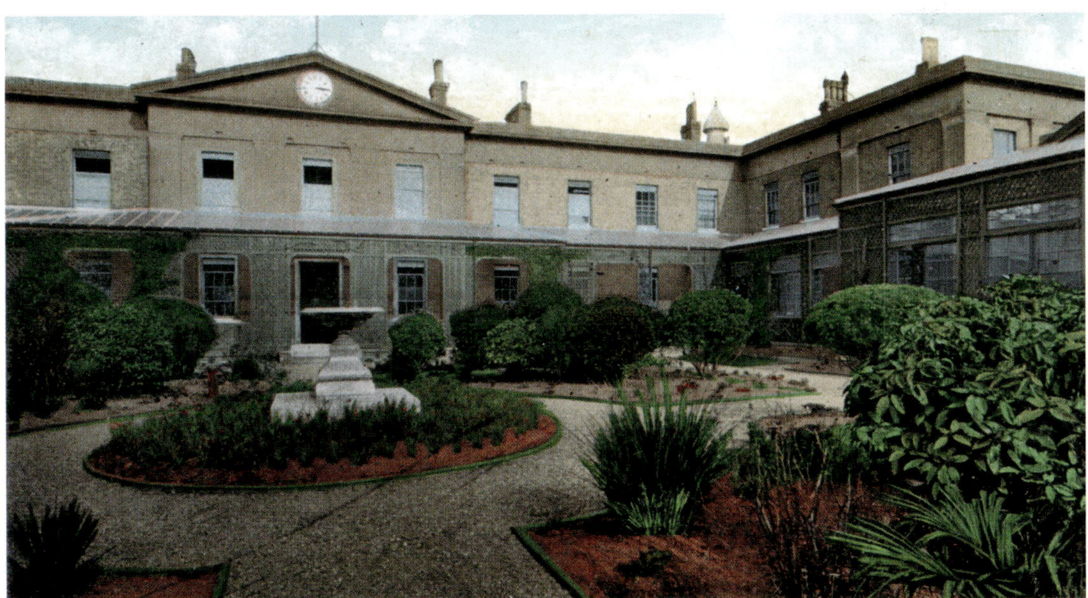

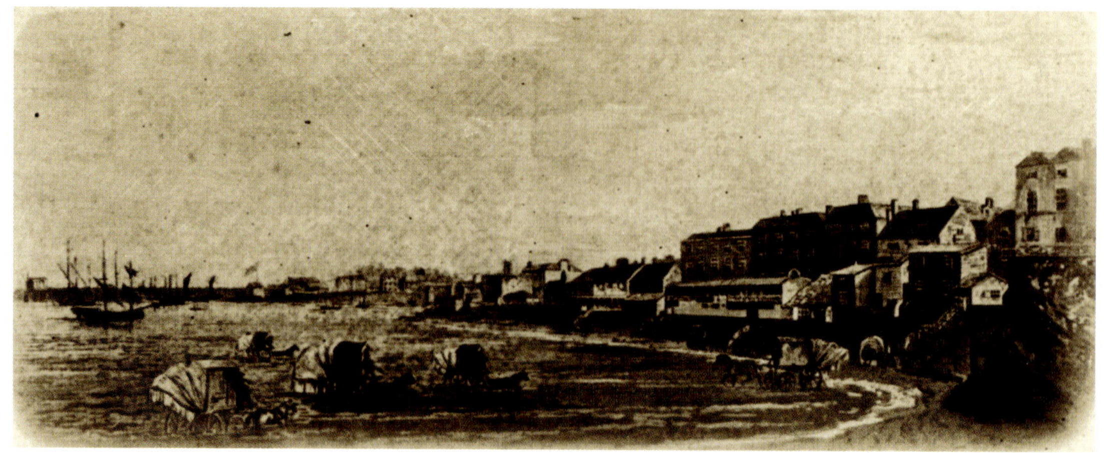

The Start of Sea Bathing

During the Georgian era, from 1714 to 1830, Margate was transformed from a small, run-down fishing village into one of our first seaside resorts. This early 1786 view of Margate beach shows a few of the horse-drawn bathing machines that were in use at the time. Margate was at the forefront of sea bathing in the eighteenth century, with bathers taken into the sea in simple carts before a fully developed bathing machine appeared there by 1753. This was ascribed to a Quaker, Benjamin Beale, who added a modesty hood to the rear of bathing machines, enabling bathers to enter the sea unobserved, and also offering some protection from wind and waves. By 1793 a guide to Margate speaks of thirty to forty bathing machines in use at a time. Bathing rooms had been established at Margate in the 1750s to the south-west of the harbour, from where bathers descended an external staircase on the seaward side of the building into a waiting bathing machine to be taken into the sea. Even the shoreline looks completely different from the present vista. Compared with the inter-war years view, it can be seen how a beach lifestyle had become popular.

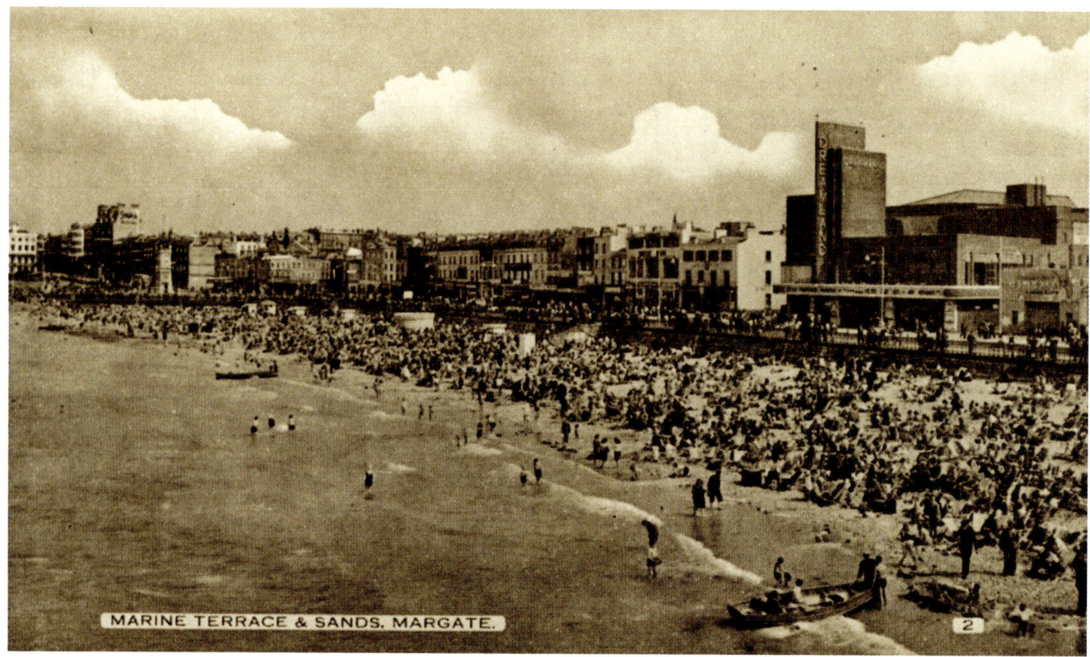

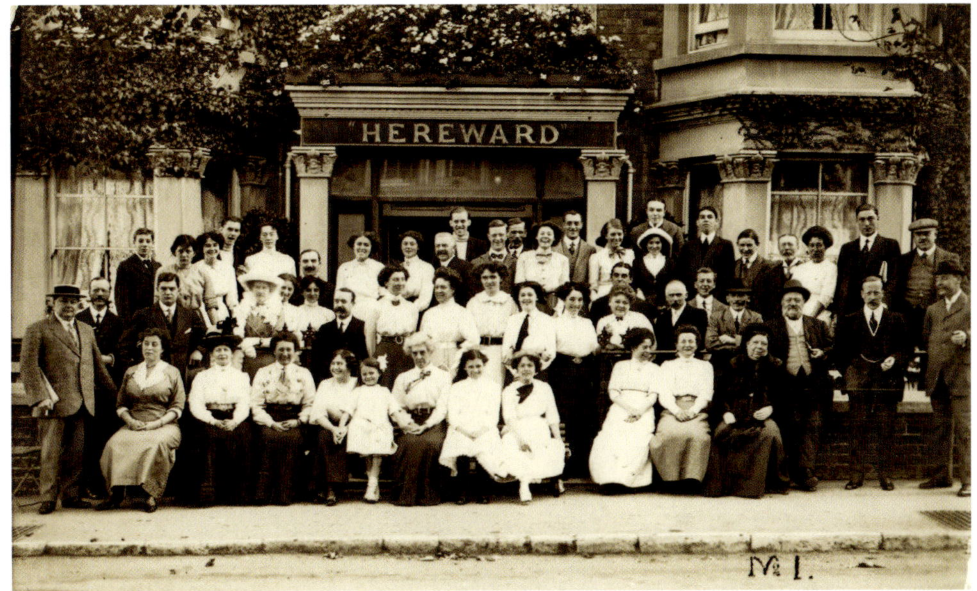

A Typical Hotel

Typical of the many hotels the new breed of holidaymakers booked into was the Hereward Hotel, which was established in around 1893 by Herbert Epps, who owned the hotel until 1923. The hotel started its life at Nos 2, 4 and 6 Gordon Road, Cliftonville, but by 1897 had grown to include Nos 8, 10, 12 and 14. In 1923 the group Hereward and Kingsley took over, adding Nos 16, 18 and 20, and so it became half of the block on the east side of Gordon Road; the Kingsley Hotel was on the adjacent Edgar Road. Here we see the guests in September 1912, compared with how the Hereward Hotel looked during resort's 1960s heyday. From the 1940s to 1980s it was numbered 2–16 Gordon Road, being converted into flats in the early 1990s and named Hereward House. Sadly, today, this area is in a very poor state.

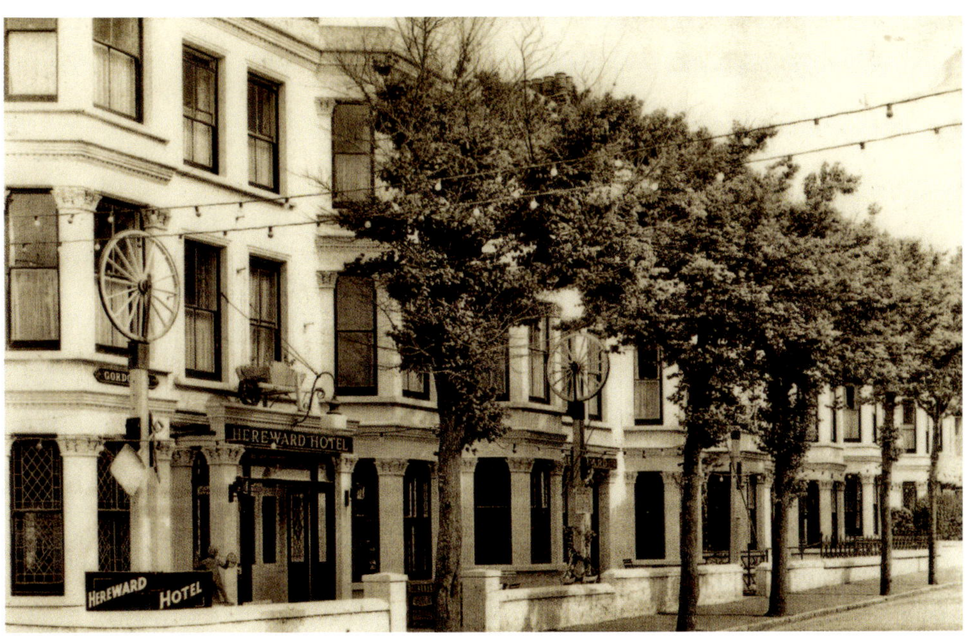

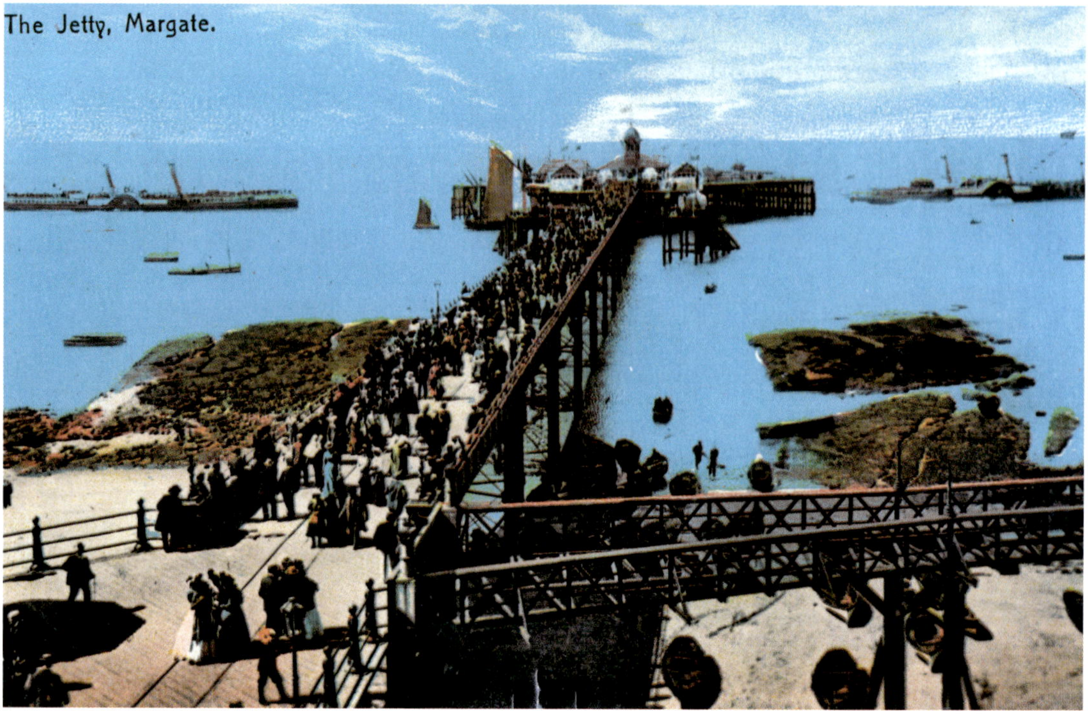

The Popularity of the Seaside

Two views of Margate's pier taken in the early 1900s, each showing how the ships used to queue near the pier-head, waiting to disembark their passengers.

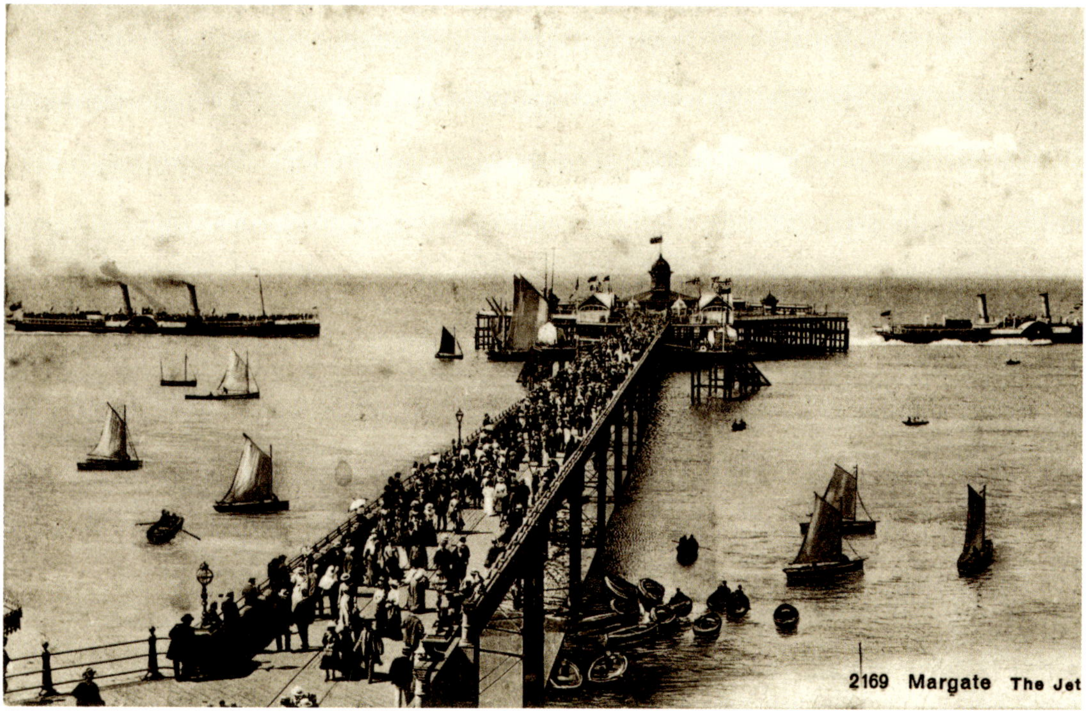

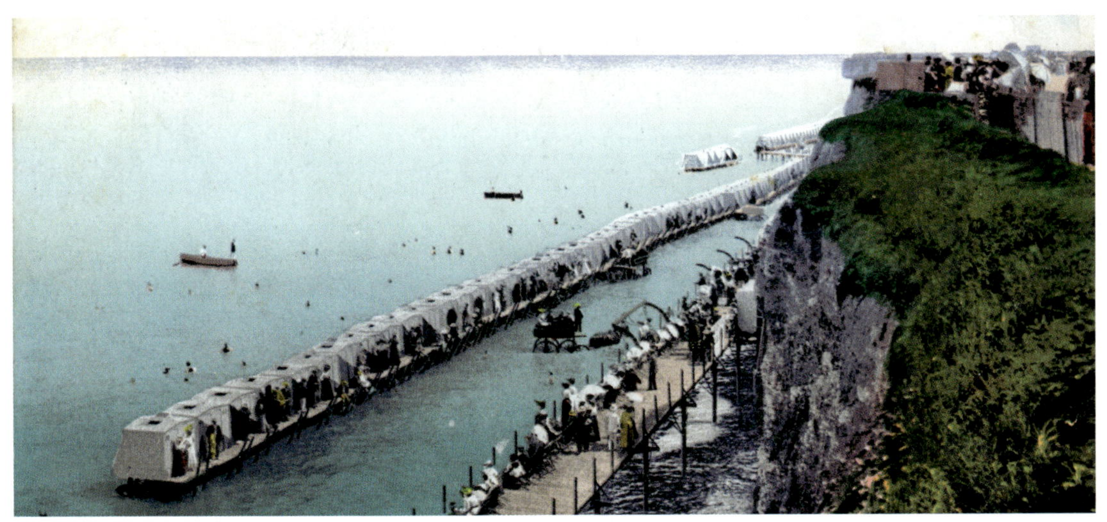

The Lido, Cliftonville

Cliftonville was once the more fashionable part of Margate, set high on the cliffs to the east of the pier and harbour. To begin with bathing was conducted from bathing machines, but in time they gave way to lidos, enclosed sea bathing pools, which became increasingly popular as the trend for sea bathing became an evermore sought-after pastime for the Victorians. The Lido at Cliftonville was built on the Clifton Baths Estate, beneath which ran a number of passageways formerly used by smugglers. This underground complex consisted of bars, cafés and an indoor, warm, sea-water pool with changing facilities. The Clifton Baths were constructed in 1824–28 by John Boys at a cost of £15,000, and were excavated from the chalk cliff to the north-east of Margate harbour. From 1926 onwards the Clifton Baths were remodelled under John Henry Iles, a leading figure in the amusement park industry between the wars; he also owned the Dreamland amusement park. The site was turned into a large, modern seaside complex with bars, cafés and restaurants on several levels and a large, open-air swimming pool projecting into the sea. These buildings were built onto and over the remaining parts of the Clifton Baths in a neoclassical style with Mediterranean influences, laid out over a series of terraces. It was renamed the Lido in 1938.

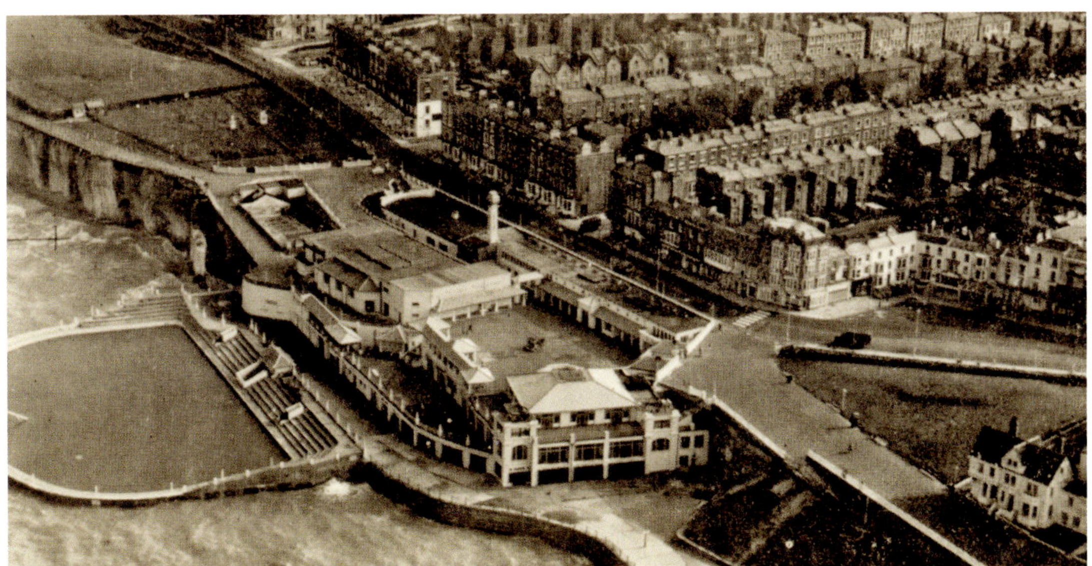

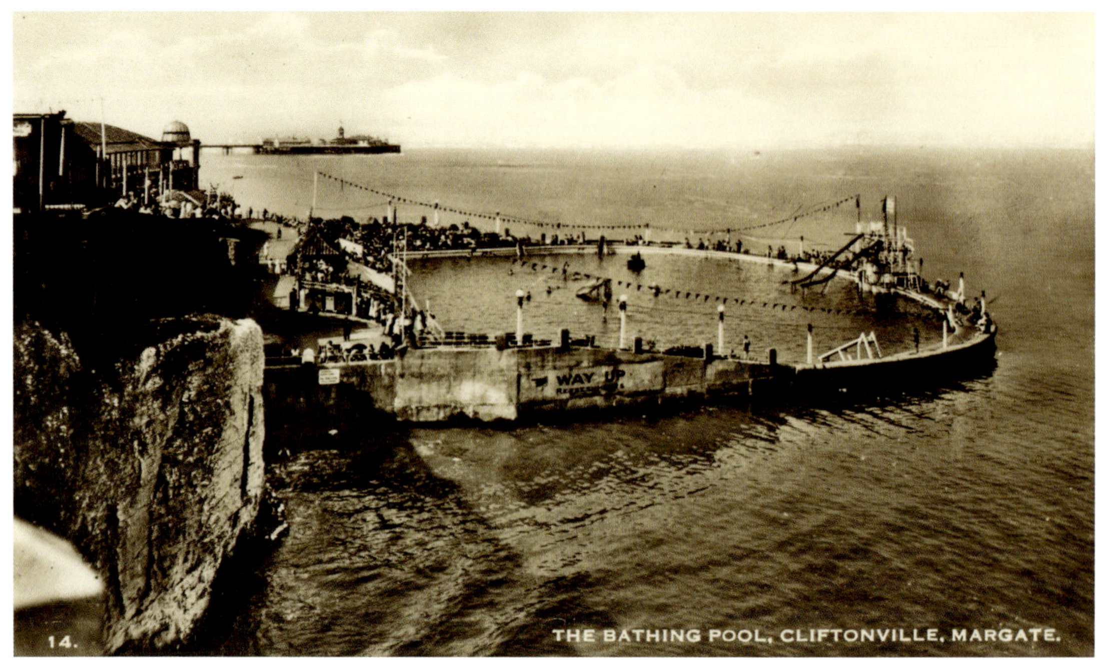

The Lido, Cliftonville

The Lido was hugely popular right through to the 1960s but a winter storm in January 1978, which destroyed the pier, also wreaked havoc with the Lido, particularly the outdoor pool. Such was the damage caused, reconstruction work was never even considered and today the Lido faces almost certain demolition. Once the centre for swimming, beauty pageants and socialising, today the site is but a pale shadow of its former glory.

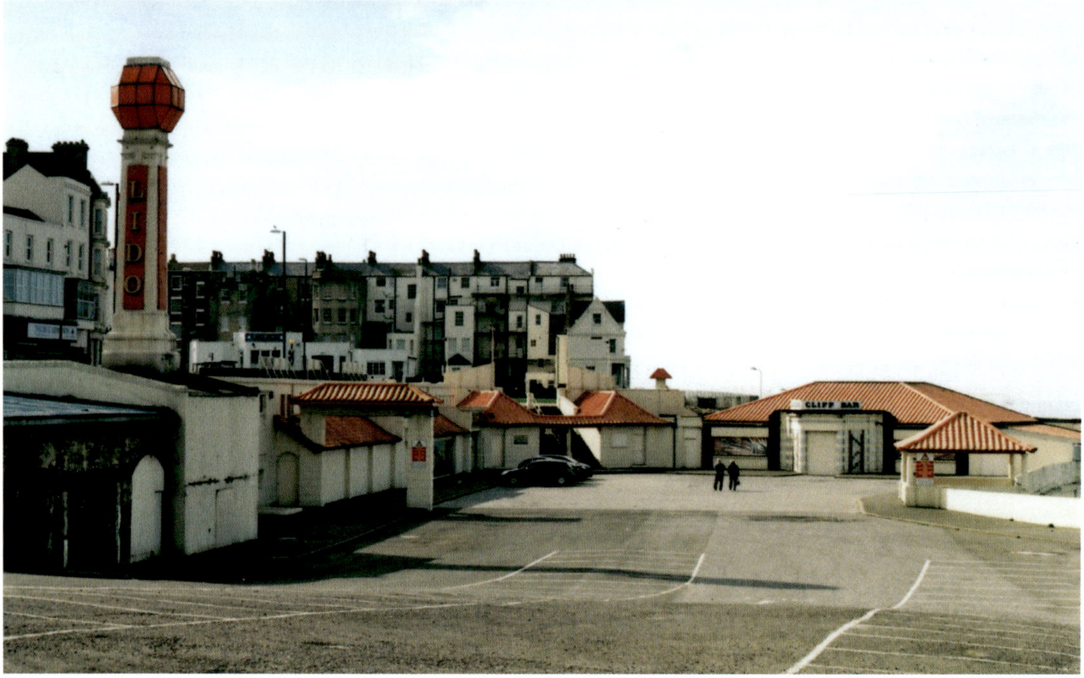

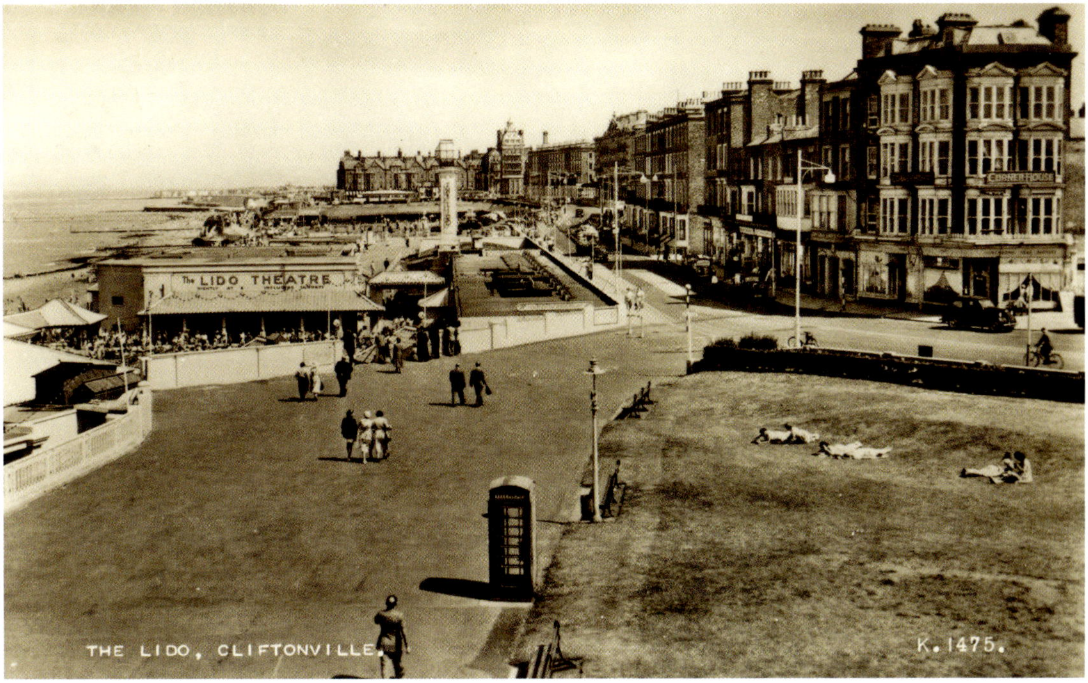

The Lido, Cliftonville

Like its near-neighbour the Winter Gardens pavilion, the Lido also had to occupy a low profile on the cliff so as not to interrupt the sea views of the existing hotels along the Parade. These two views, separated by twenty years in time, show how unaltered the complex remained.

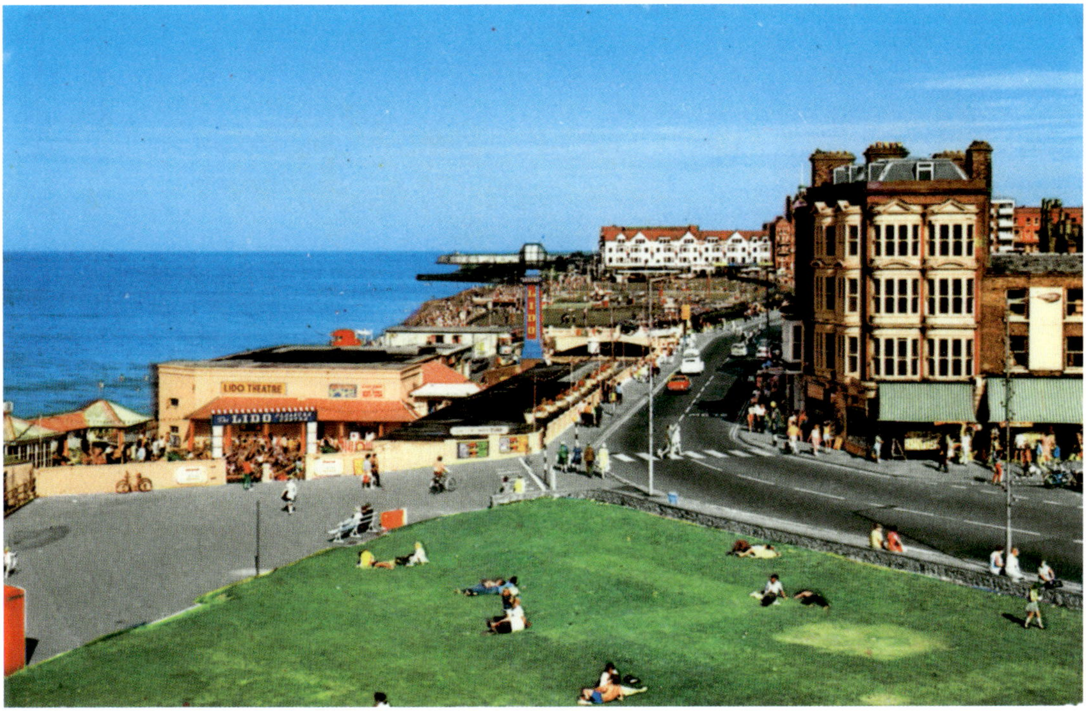

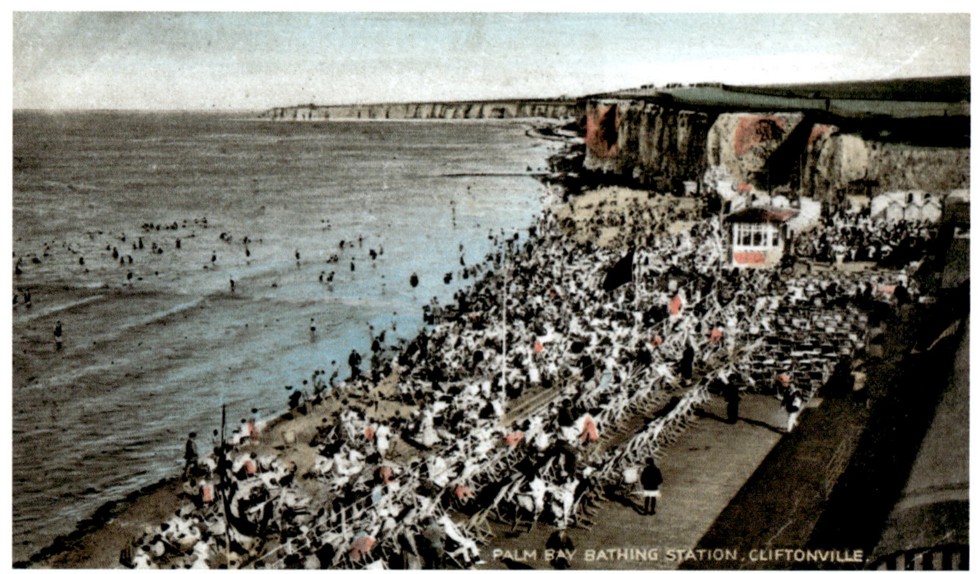

Palm Bay, Cliftonville

The original Palm Bay Estate was built in the 1930s as a number of large, wide avenues, like Gloucester Avenue and Leicester Avenue, with detached and semi-detached houses with driveways, garages and gardens. This land had been sold by Mr Sidney Simon Van Den Bergh to the Palm Bay Estate on 23 June 1924. The estate covers the eastern part of Cliftonville and was fields when the first house was built. It extends east beyond Northumberland Avenue and has been developed in phases. An earlier phase covered the northern ends of Leicester and Gloucester Avenues, and the whole of Clarence and Magnolia Avenues, while the later phase, extending eastwards from Princess Margaret Avenue, is a Wimpey-style housing estate with small houses, largely identical in appearance and of less substantial build-quality than the original 1930s estate. The beach at Palm Bay became especially popular with holidaymakers and day-trippers alike. Today it is less hectic as a lone sunbather looks pensively out to sea.

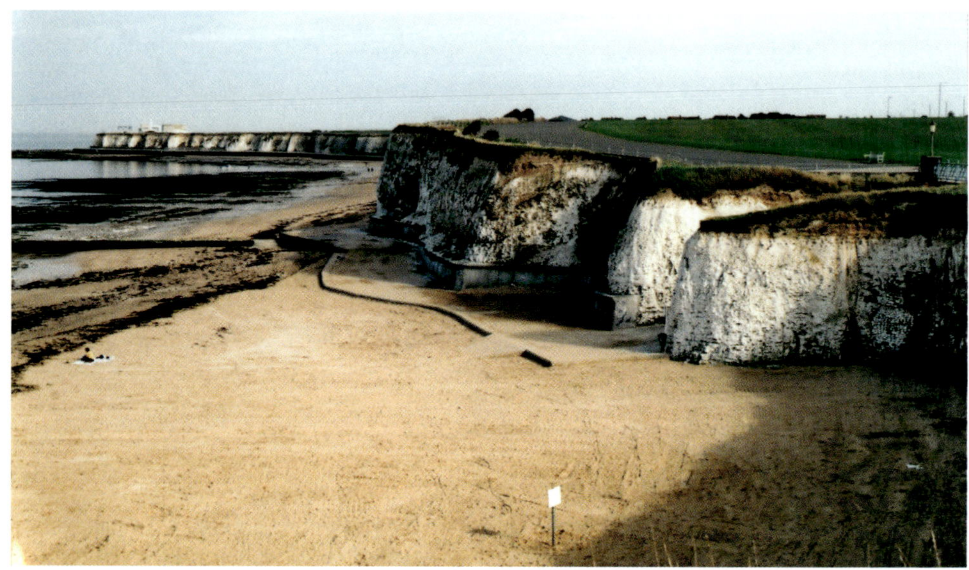

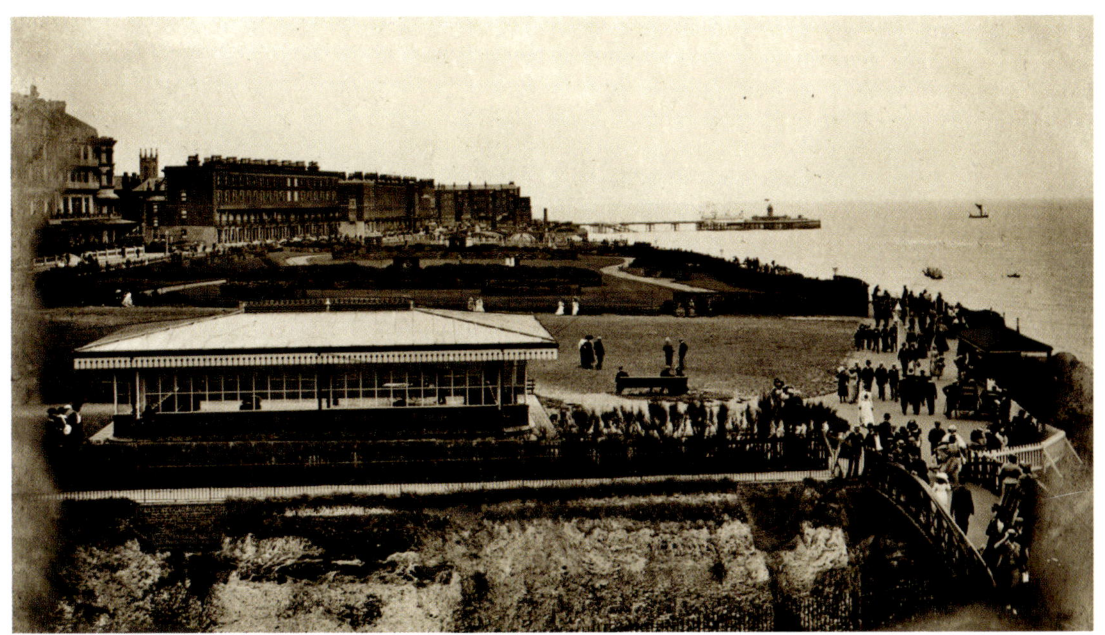

Clifftop Gardens

For our Victorian and Edwardian ancestors there was nothing quite like a stroll through gardens such as these. In the background can be seen Margate Pier. The bridge seen in the bottom-right corner of the old view is that which runs across the centre of the modern-day view. It was not possible to work out exactly where the old-time photographer stood to capture his view.

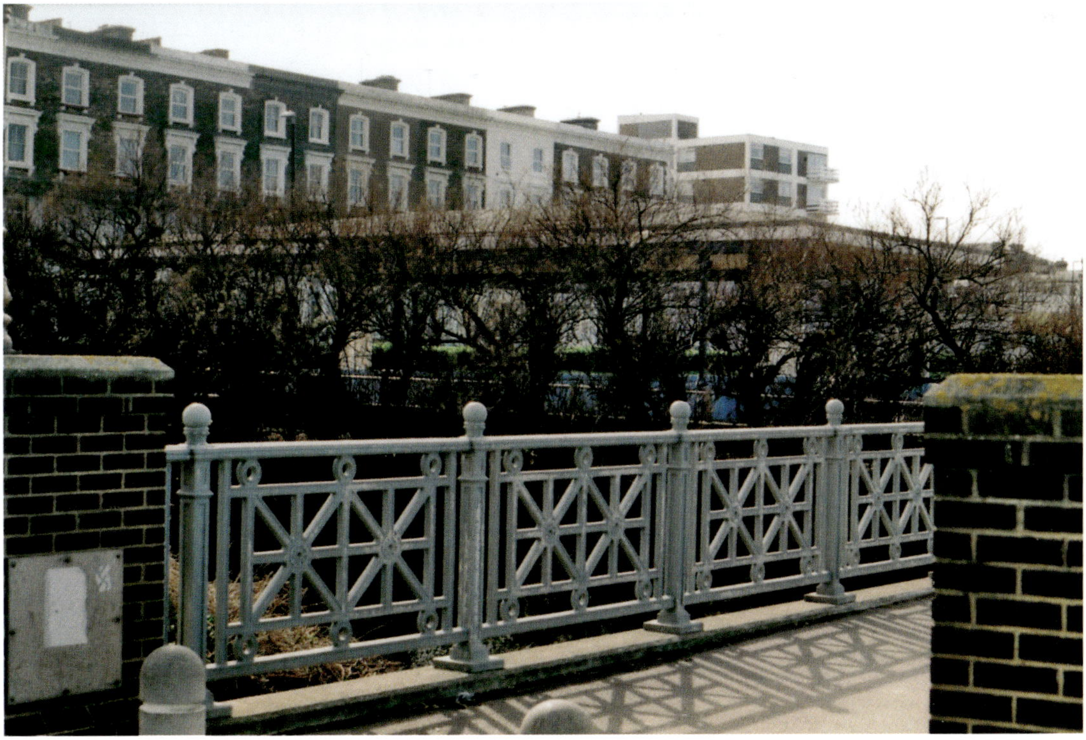

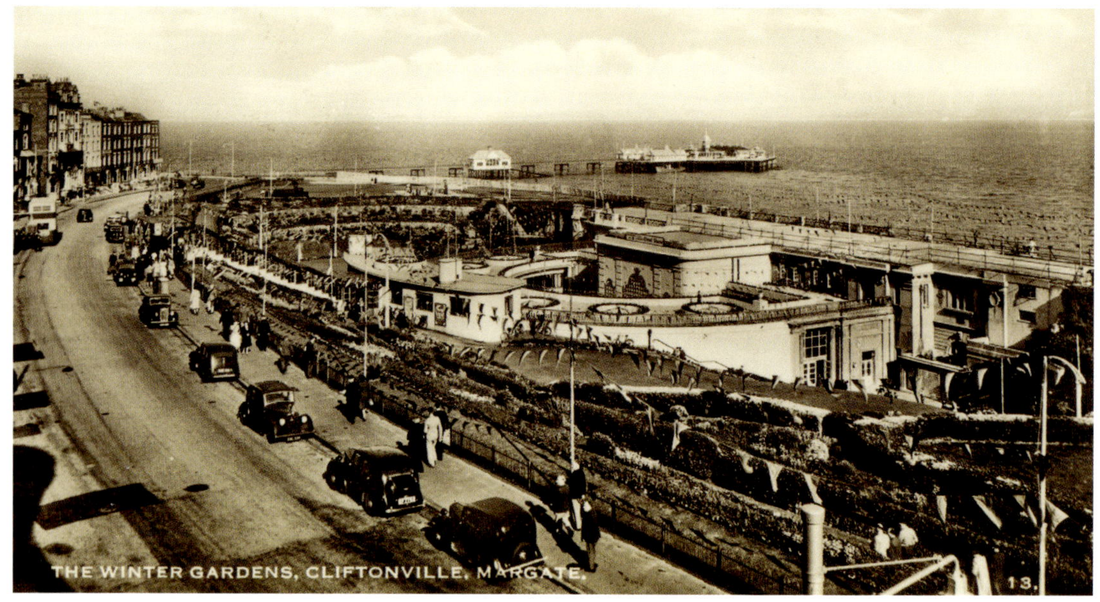

The Winter Gardens

The Winter Gardens was opened on 3 August 1911 by the then-mayor of Margate, Alderman W. B. Reeve; it took just nine months to build at a cost of £26,000. The history of municipal entertainment in this country can be traced back to Margate, which blazed the trail that others were to follow. The main reason for the Winter Gardens being built in an artificial hollow was that the existing buildings around Fort Green had a covenant which did not allow the erection of any building on the Green that could obscure the view or light of the ground floors of these buildings. When completed, the Winter Gardens consisted of a large concert hall, four entrance halls, two side wings and an amphitheatre. Internally, the Winter Gardens was decorated in a Neo-Grecian style. Originally the stage could be viewed from both the main hall and the amphitheatre, with the ability to enclose the stage in bad weather. The accommodation was for about 2,500 people inside the building and 2,000 in the open air.

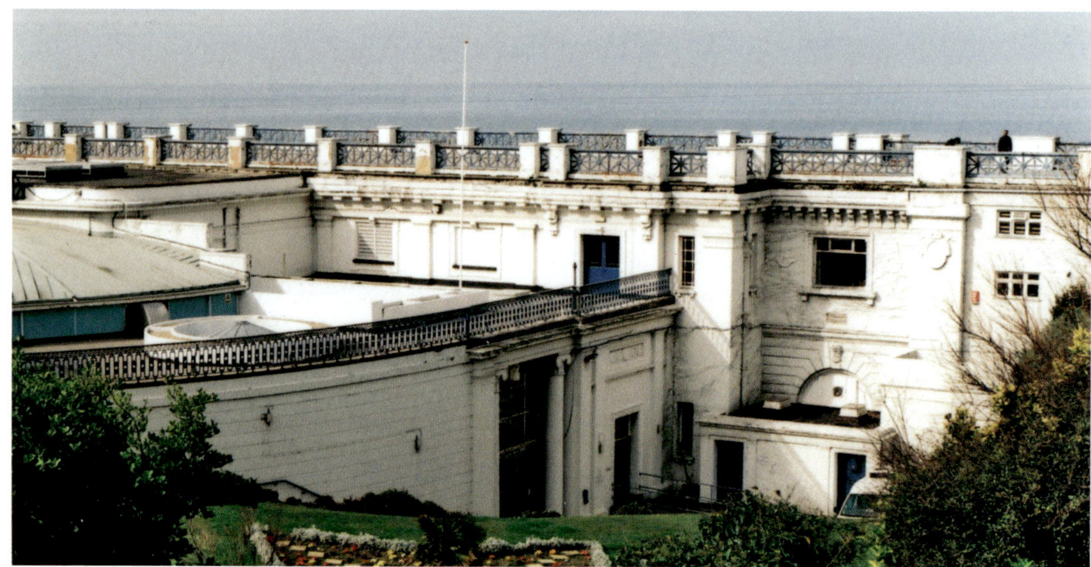

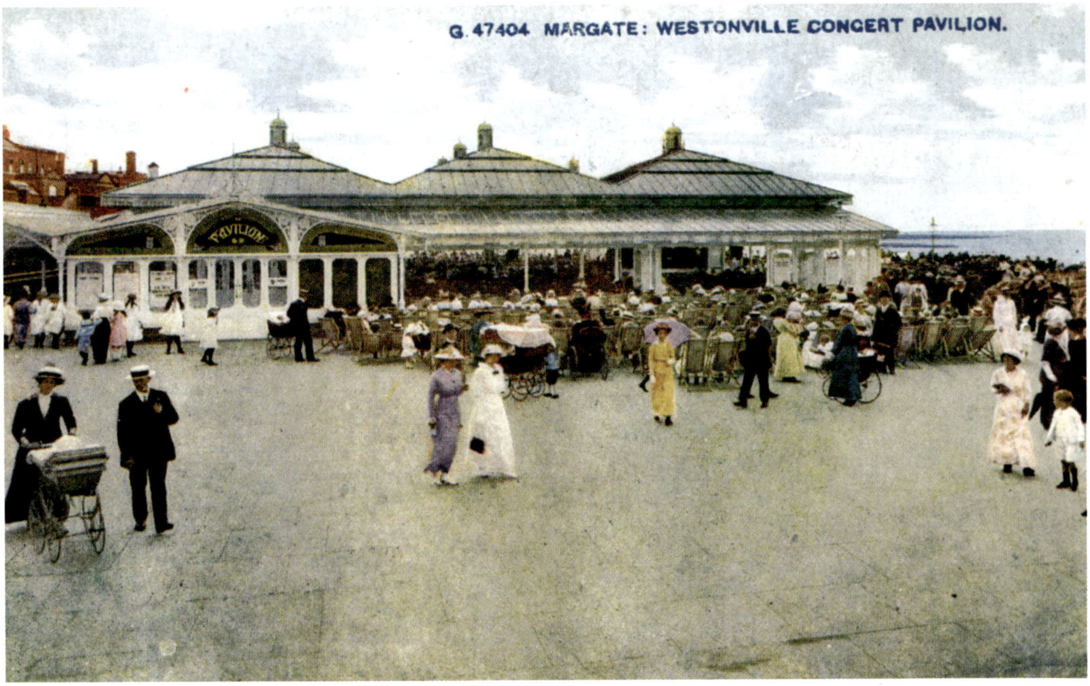

Westbrook Pavilion

At the other end of the municipal entertainments scale was the Westbrook Pavilion, situated on the other side of Margate at Westbrook. Built sometime during the Victorian/Edwardian period, it regularly entertained holidaymakers until it was swept away by the East Coast floods of 1953, after which the site stood derelict for a number of years until it was turned into a children's amusement park.

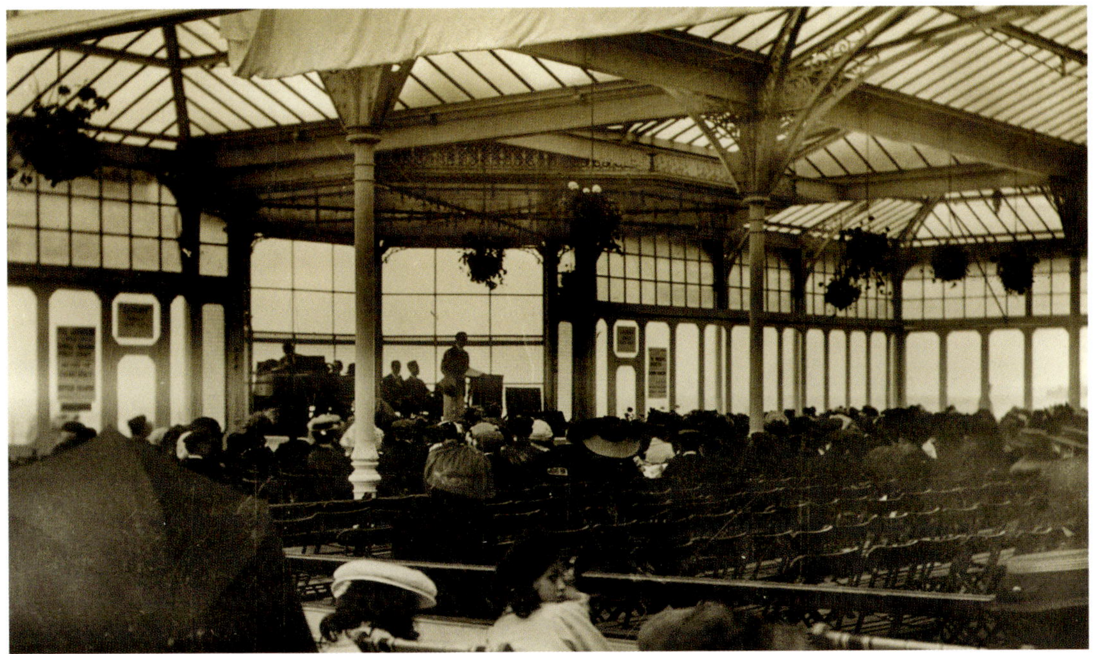

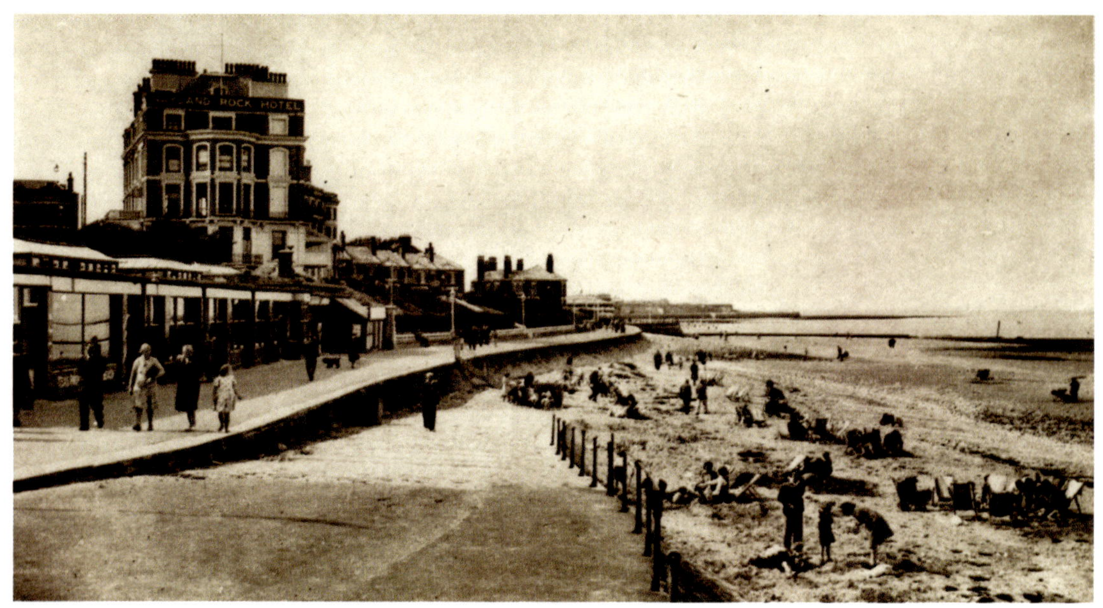

Westbrook Sands and the Nayland Rock

Known by many day-trippers and holidaymakers alike as 'the quiet end of Margate beach', this area was eminently suitable for those who wanted a quiet day on the beach with few distractions. It was ideal for those with youngsters who simply wanted to play on the sand or explore the many rock pools to be found on the Nayland Rock at low tide. Ahead, just beyond the final block of buildings can be seen the Westbrook Pavilion. In the modern-day view the derelict site to the left is all that remains of the Nayland Rock Café and the adjoining shop, both much used by my family when we were growing up.

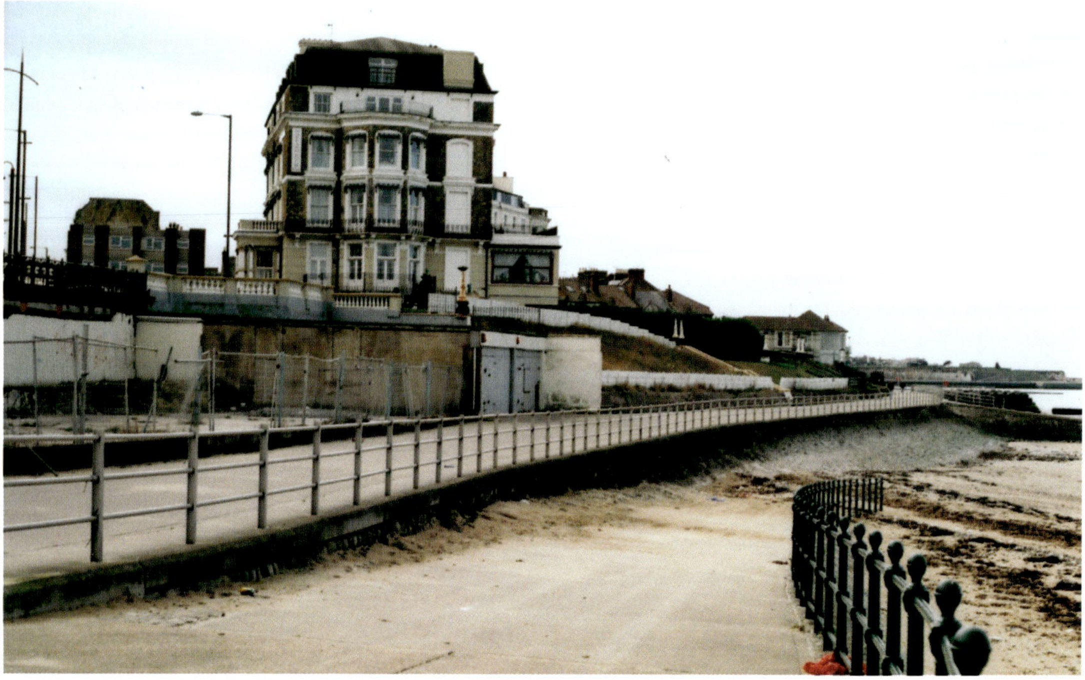

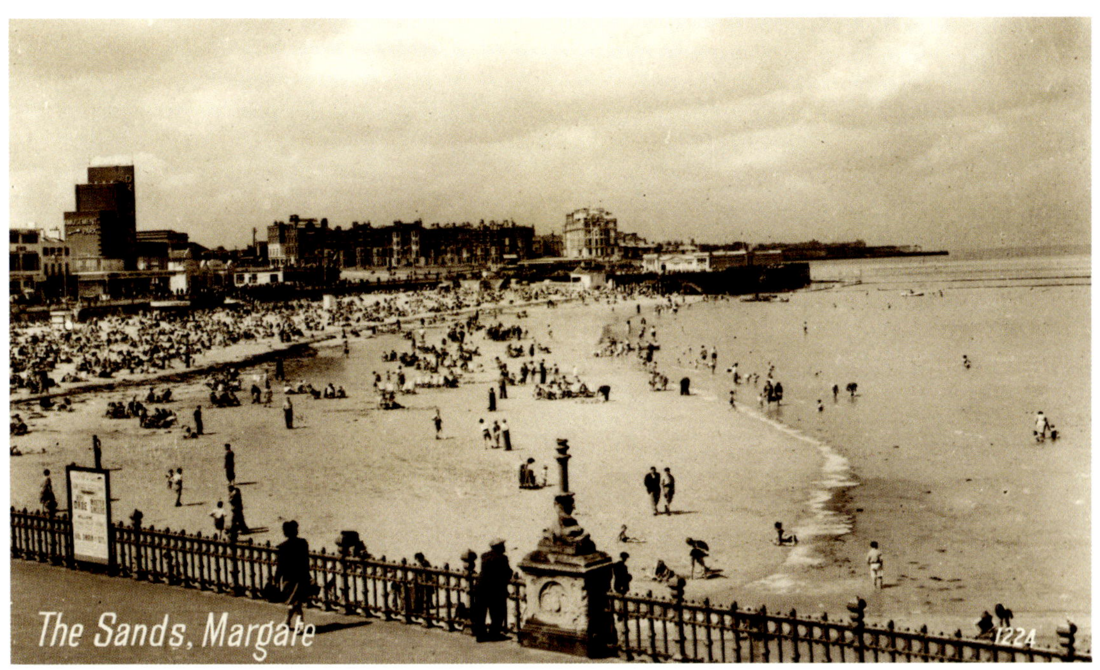

Margate's Main Beach

Looking westwards across Margate beach you get an idea of how compact it is. The raised-platform structure in the top view shows the Sun Deck, which was opened on 23 July 1926. In front of it at low tide was a swimming pool, built to retain water from the receding tide; it accommodated 320 bathers. The Sun Deck was a smaller version of the much grander lidos that many resorts, including Margate, had at this time. This construction survived numerous storms in its time and a World War, before being demolished in the late 1980s.

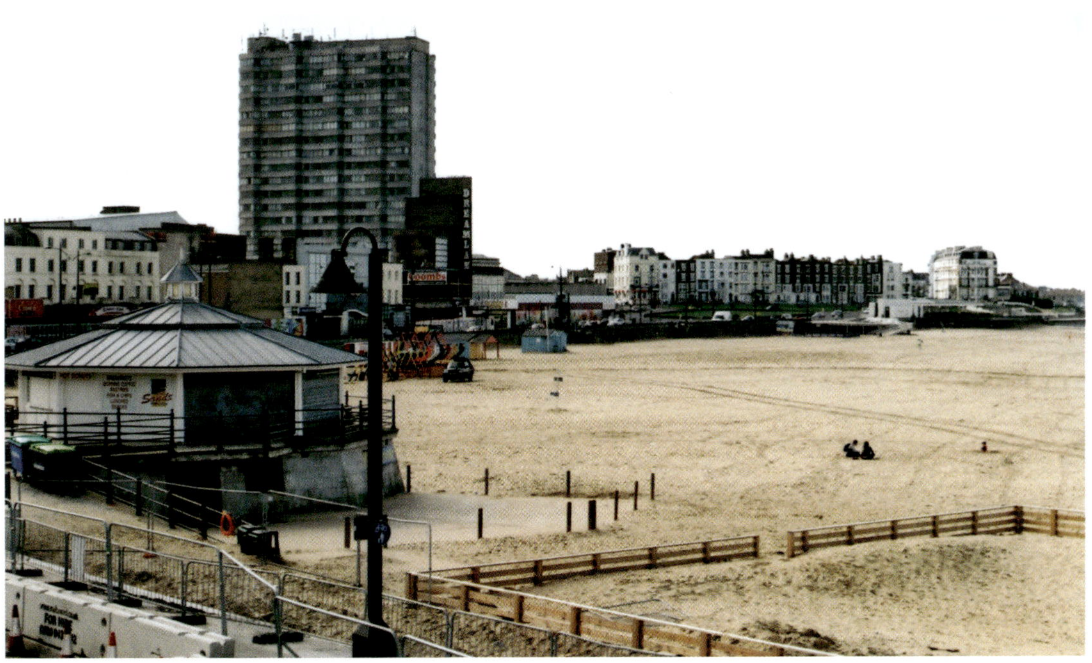

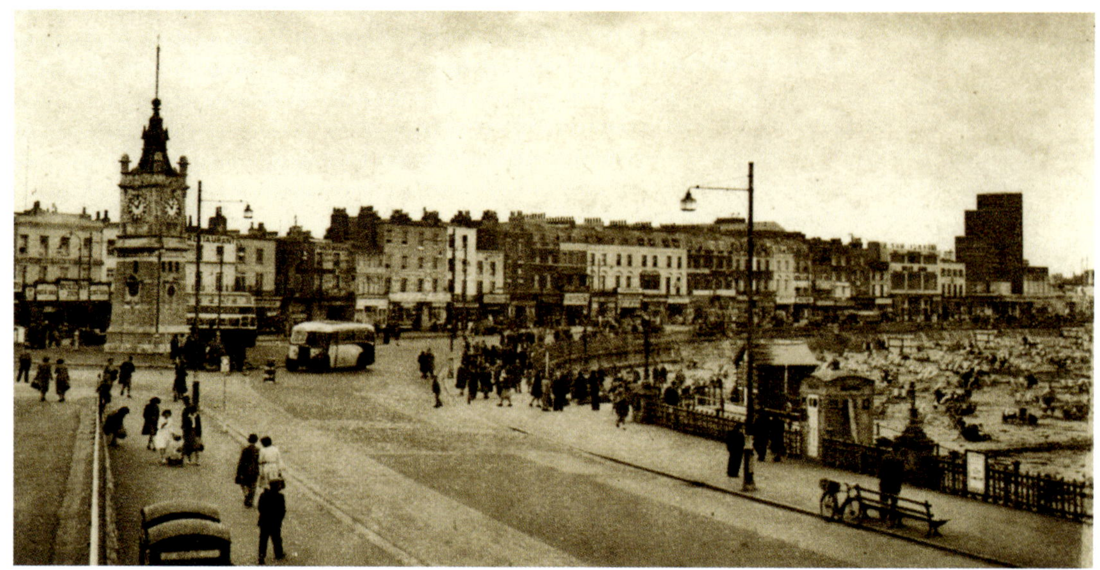

An Archetypal Seafront

What made Margate the holiday resort it became in the post-war years was its seafront, with a variety of amusement arcades, gift shops, cafés and restaurants. I can well remember, as a small boy on holiday at Margate, spending many a happy evening with the rest of my family just walking up and down this area with its heady mix of the smells of candy floss and ice cream, and the sounds emanating from the amusement arcades. Back then the vista was dominated by the Dreamland tower, but in around the 1960s a hideously tall block of flats, Arlington House, was built and has since become a blot on the streetscape. It is long-overdue for redevelopment.

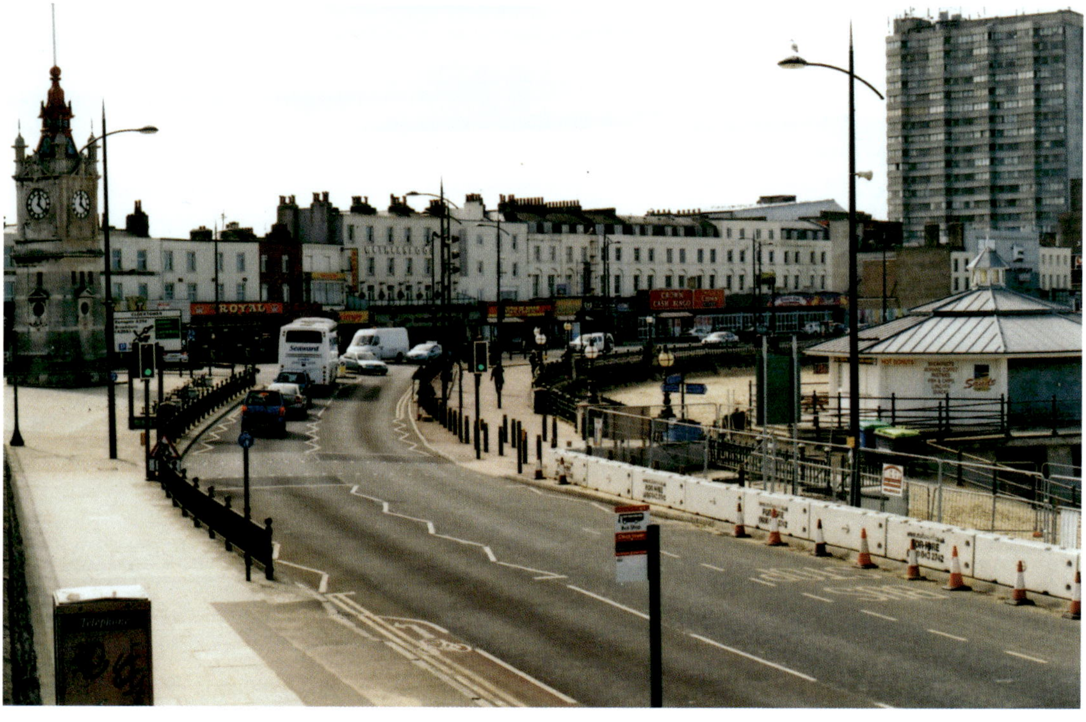

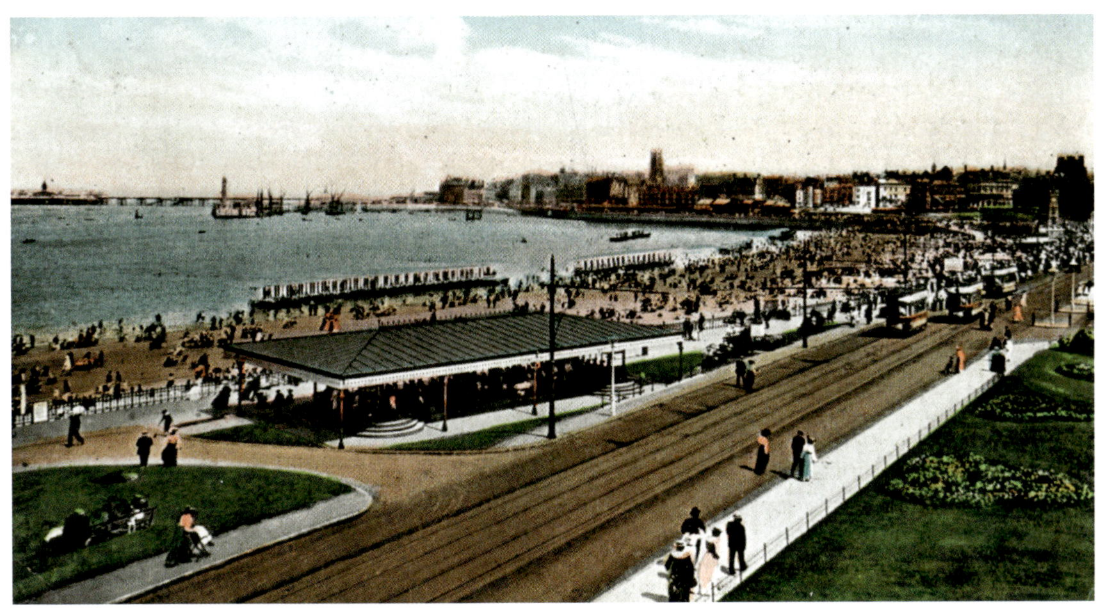

Margate, Looking Eastward

As can be seen, in 1906 bathing machines could still be had on the main beach and sunbathing was becoming increasingly popular, judging by the crowded beach. The shelter to the left of both pictures is of particular interest. Apart from it being a listed building, it was here, it is said, the poet T. S. Eliot sat to compose the key lines of his poem, *The Waste Land*, which includes the line, 'On Margate Sands I can connect nothing with nothing.' Yesterday's view of the harbour is now blighted by the Turner Contemporary art gallery, which was opened in April 2011.

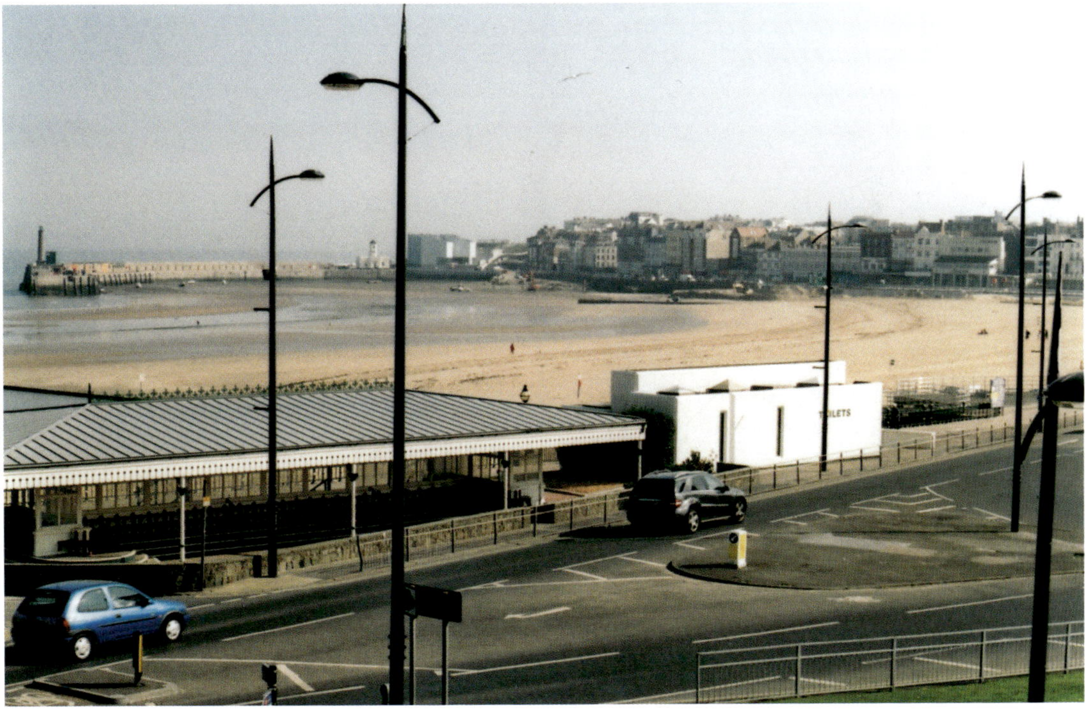

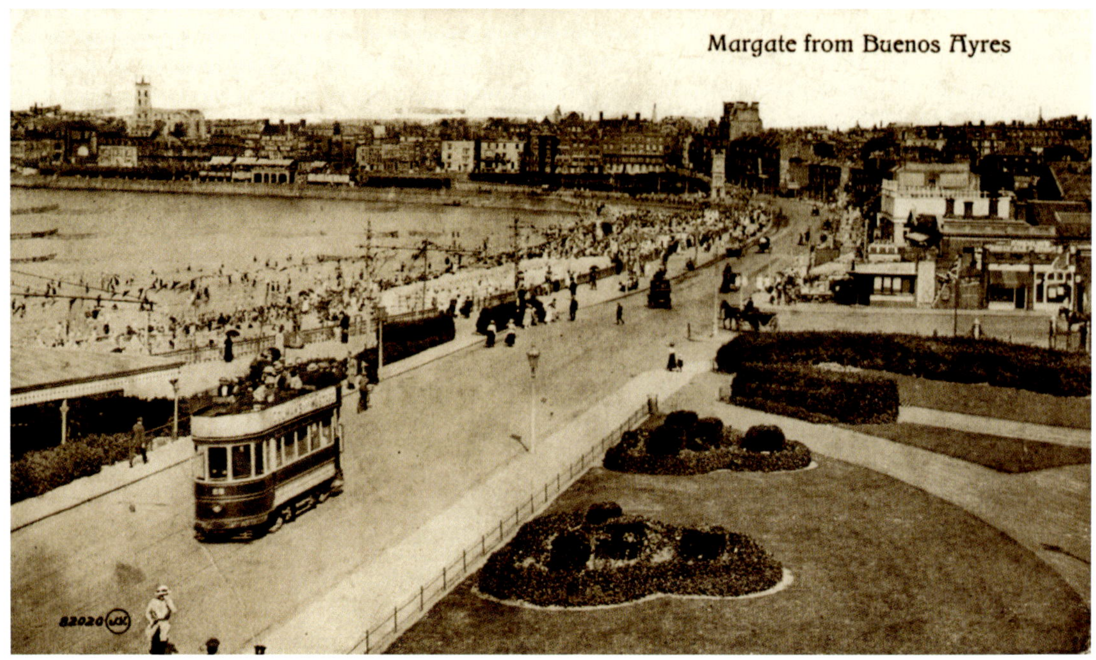

The Seafront

The Buenos Ayres is a guest-house that overlooks the seafront, close by the Nayland Rock. This top view was taken in 1923 and as can be seen, the view from the first-floor balcony is remarkably unchanged over the intervening ninety-odd years. Only the Dreamland tower and Arlington House have been added to the vista. Also, originally the public conveniences were below ground, out of sight, whereas today they are almost celebrated by a white-painted box-like structure. This was a guest-house where the Clancy family stayed in the late 1940s.

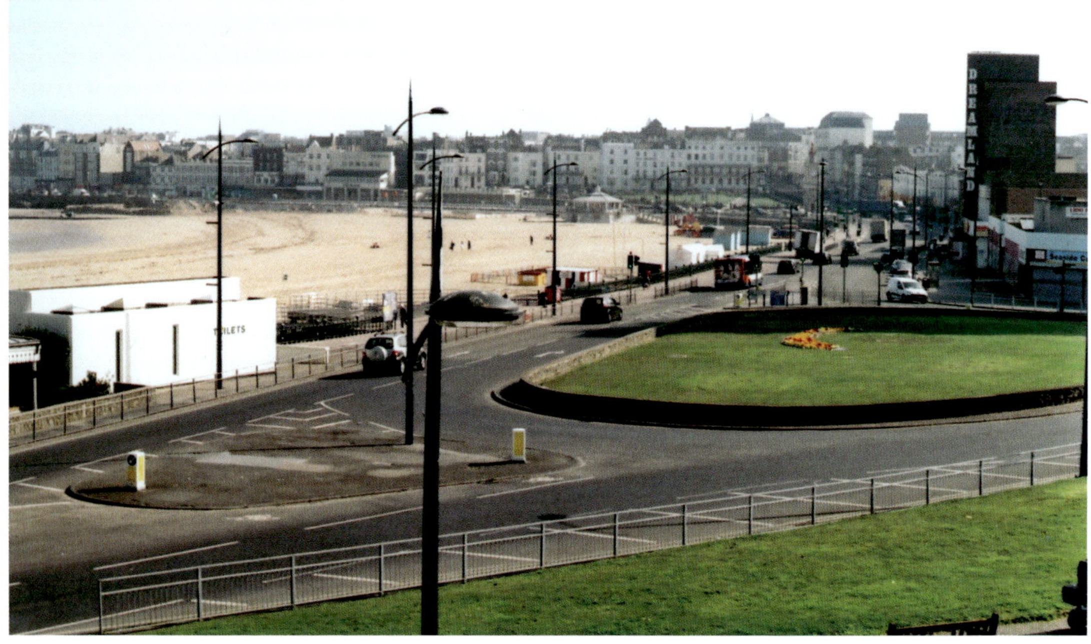

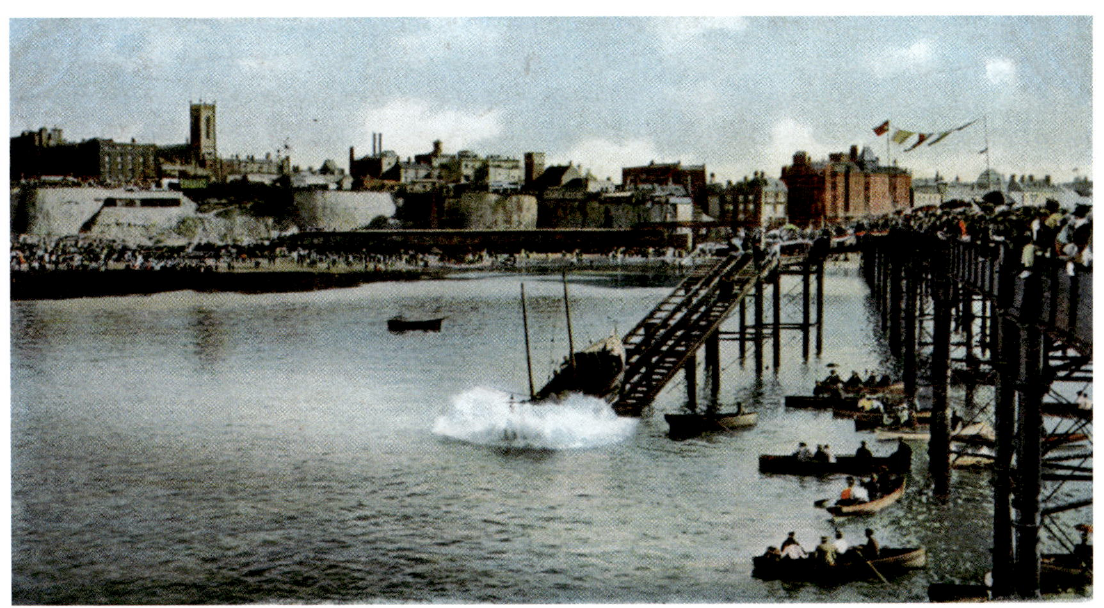

The Lifeboat

An endearing part of seaside life to many is the lifeboat. Most holidaymakers cannot resist visiting the lifeboat station, and many of these were to be found on the pier. Margate's lifeboat was stationed on the pier, at the top of a ramp for a speedy launch. I well remember, as a small boy playing on the sands, I would listen out for the distinctive 'crack' of the maroon being fired to summon the lifeboatmen to the lifeboat station, then after waiting for a few minutes you would start to see men racing along the pier. These were days prior to pagers and mobile phones. Following a storm in January 1978, much of Margate Pier was destroyed, the lifeboat station became inaccessible and a new home had to be found. It can be seen in this 1906 view of the lifeboat – possibly the *Quiver* – being launched that at that time there was no boat house in which to keep the vessel.

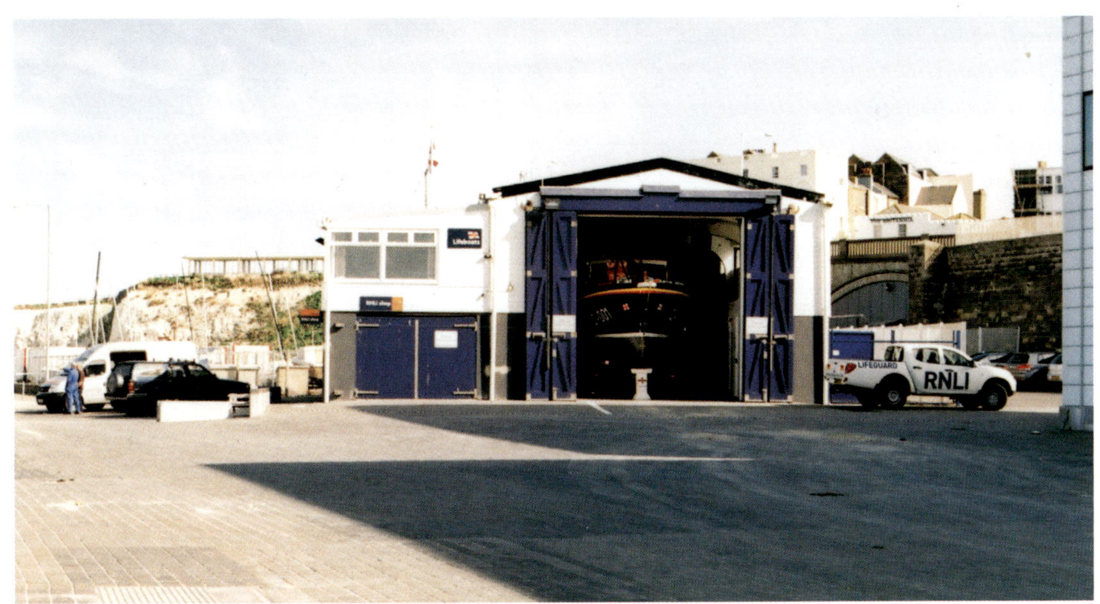

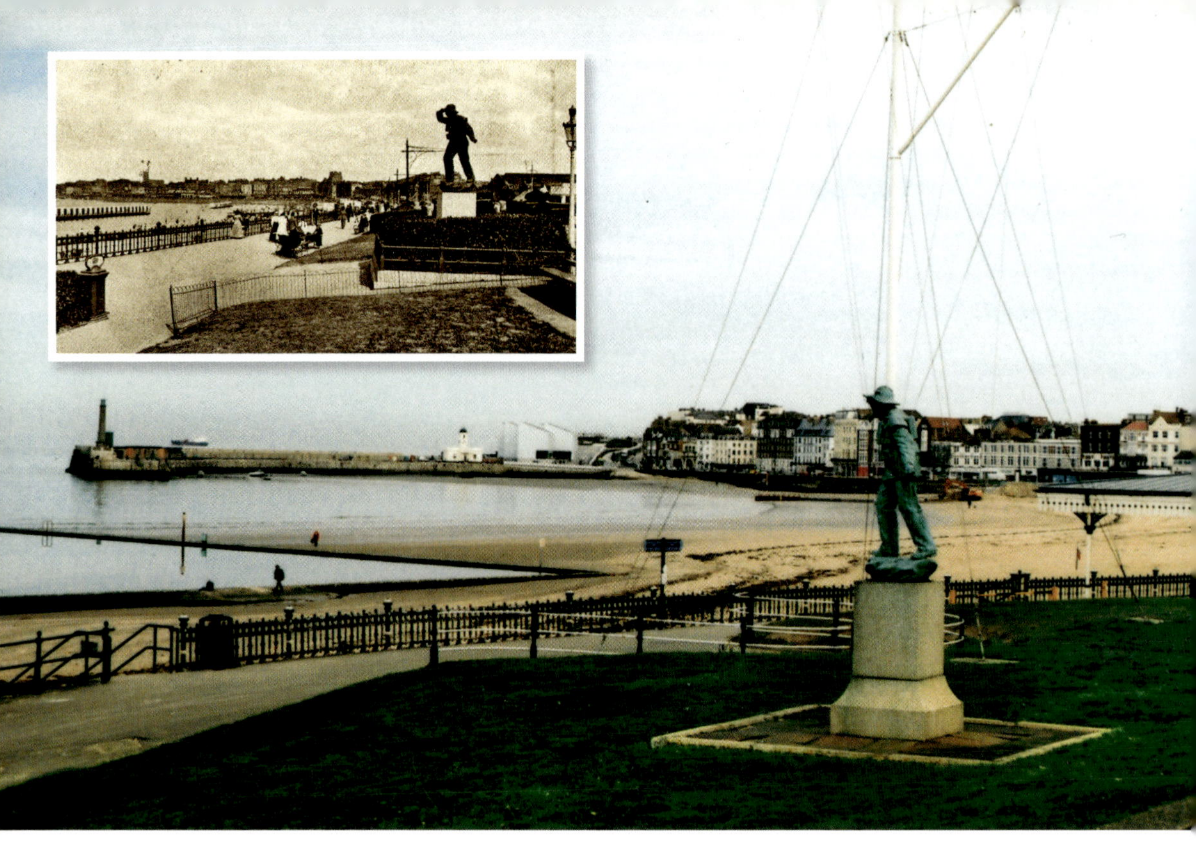

The Margate Surf-Boat Memorial

When entering Margate seafront from Westbrook, it is impossible not to notice the statue of a figure looking out to sea over the Nayland Rocks. The figure is shading his eyes and is clad in oilskins and an old cork life-jacket of yesteryear. This monument was erected to the memory of nine Margate men who lost their lives in the *Friend to all Nations* surf-boat tragedy on the night of Thursday 2 December 1897. On that fateful night a great storm grew to cyclonic proportions and all but destroyed the town, its harbour, seafront and main roads. Conditions that night could not be matched to any other natural disaster known in local history or living memory. The storm raged throughout the night, yet at daybreak the men of the surf-boat, *Friend to all Nations*, undaunted, responded to a distress flare fired from *The Persian Empire* off Margate. Surf-boats can be likened to today's in-shore rescue boats, supplementary to the main RNLI lifeboat. Margate's surf-boat, like most others, had no motor, was powered by oars and sails, and was not self-righting. Thirteen boatmen on the surf-boat launched at 5:20 a.m., a full ten minutes ahead of the larger Margate RNLI lifeboat *Quiver*. Unfortunately the boatmen were caught by a sudden squall, causing serious listing under the weight and momentum of the sea water. The boat, not having time to right itself, was then hit by a further wave causing the *Friend* to turn over onto its keel, trapping one man, Joe Epps, inside and throwing those that had not jumped with all their strength, including men that had been knocked unconscious in the boat, into the freezing water, beneath which hid many jagged and seaweed-covered rocks. The capsized surf-boat later drifted ashore and beached on the tide; it came to rest by the Nayland Rock. Of the thirteen who put to sea that morning, there were just four survivors. A disaster fund was set up locally and sponsored nationally by *The Daily Telegraph* for the widows and orphans of the lost boatmen, although most belonged to the Ancient Order of Oddfellows or one of the Friendly Societies for support in exceptional times. Subsequent publicity provided some £10,000, money which was mostly spent on two large memorials, one in the local cemetery and the other on the Parade overlooking Nayland Rock, where the tragedy took place.

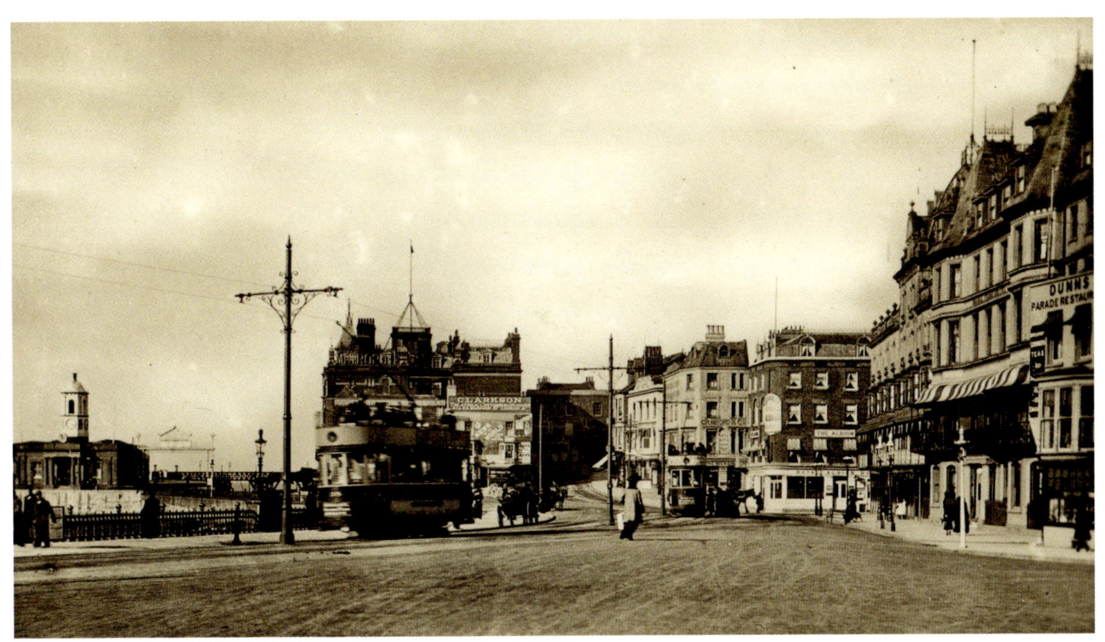

The Old Town

Margate as a town developed from the area beside the harbour and to this day there are many delightful original Georgian buildings still to be seen here. In 1907, it was a quiet thoroughfare with just the occasional tram passing through. Today there is much more traffic to contend with.

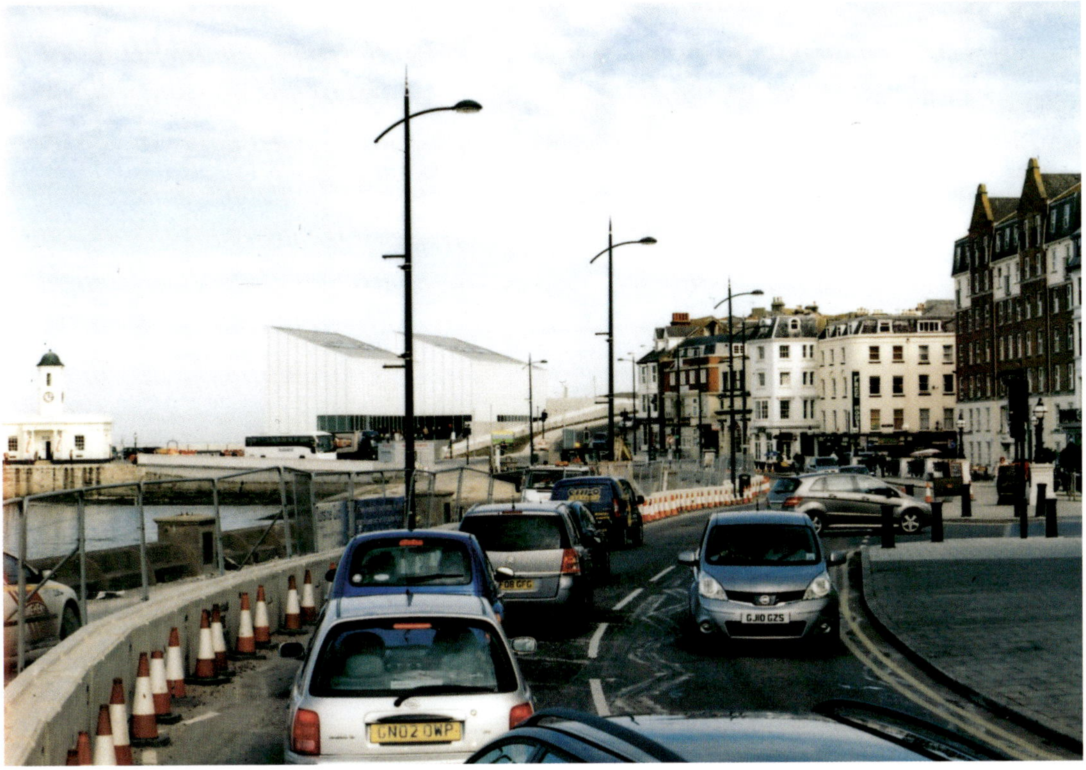

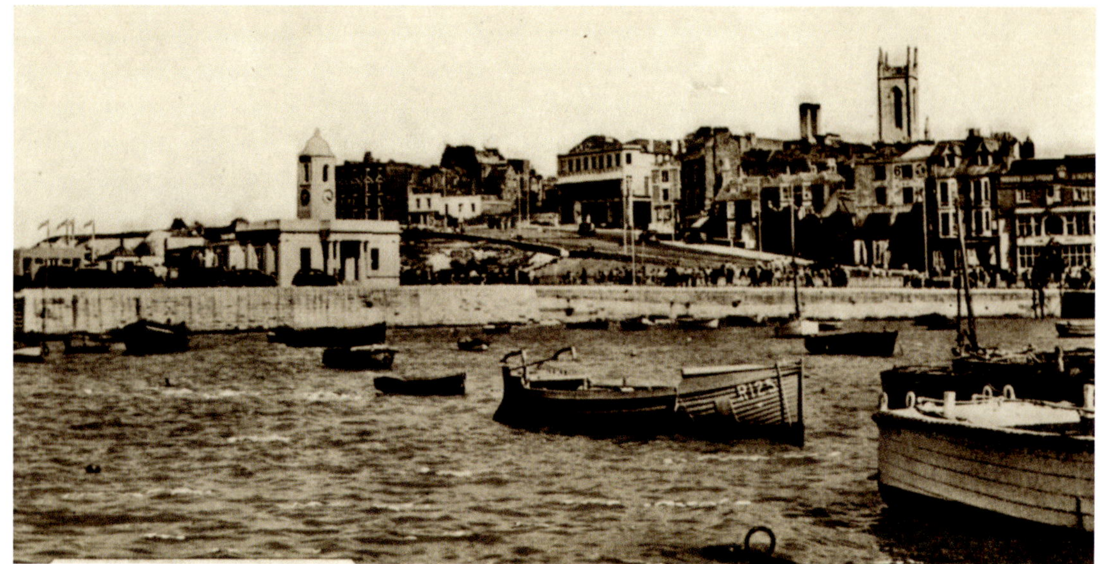

The Harbour

Confusingly, the stone jetty that forms the backdrop to the harbour was originally known as Margate Pier, and what later became known as the pier was in fact Margate Jetty. It was built between 1810 and 1815 by John Rennie so that steamships could use Margate and it became known as the Harbour Arm. The 'pier' came later, in 1855, and was built by Eugenius Birch. In the centre of both views, but now seriously overshadowed by the Turner Contemporary art gallery, is Droit Office, which was constructed in 1828 as an office for the collection of harbour dues by the Pier and Harbour Company. Maps, engravings and photographs indicate that the structure was originally rectangular in plan and the rear of the office faced directly out to sea on the outer harbour wall. The wood-and-iron jetty, constructed between 1853 and 1856 by Birch, extended to the rear of Droit Office, so extensions were added to the rear of the original Droit Office between 1852 and 1872. In the 1940s the Droit Office was destroyed by German bombing, and was subsequently rebuilt in 1947 on the original footprint according to the original architects' plans. The building extensions to the rear of the Droit Office do not appear to be replaced at this time and a separate rectangular building was constructed on the Jetty entrance. The current Droit Office has an extension to the rear constructed at the end of the twentieth century and now serves as the visitors' centre for the Turner Contemporary.

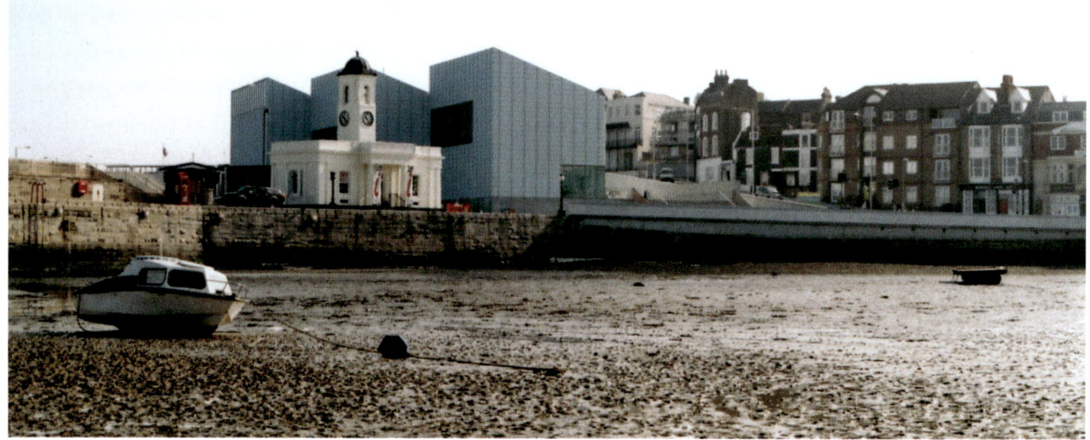

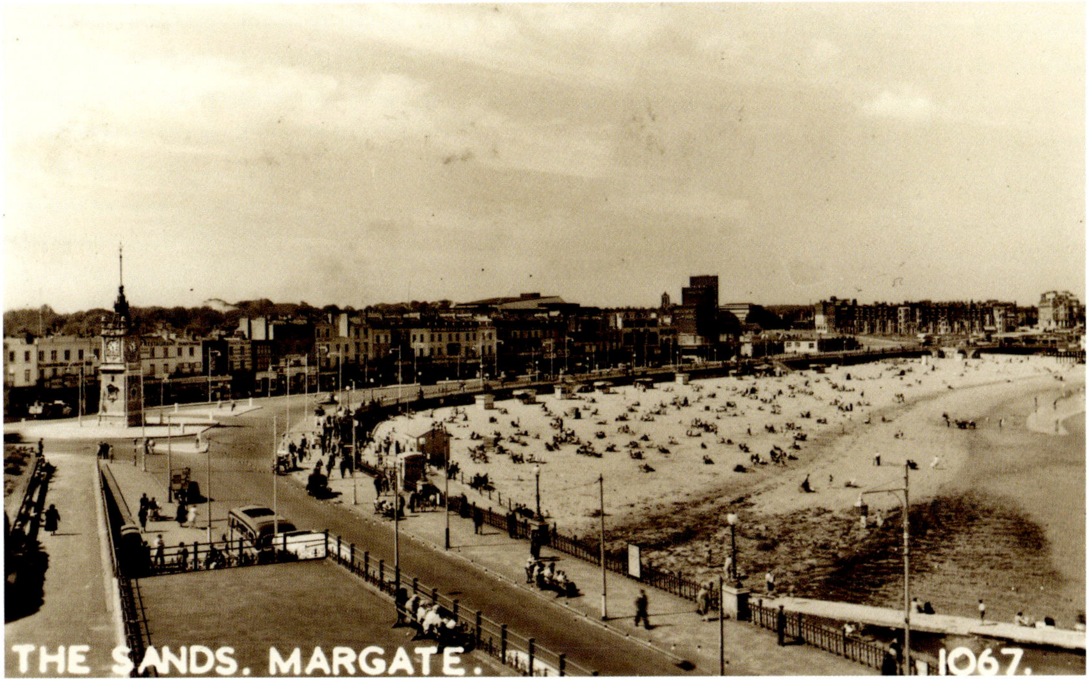

Margate Sands

Still one of the finest beaches in the south-east, Margate has all the necessary ingredients for a successful seaside resort, as can be seen in this view taken in 1962 when the resort was at the height of its popularity.

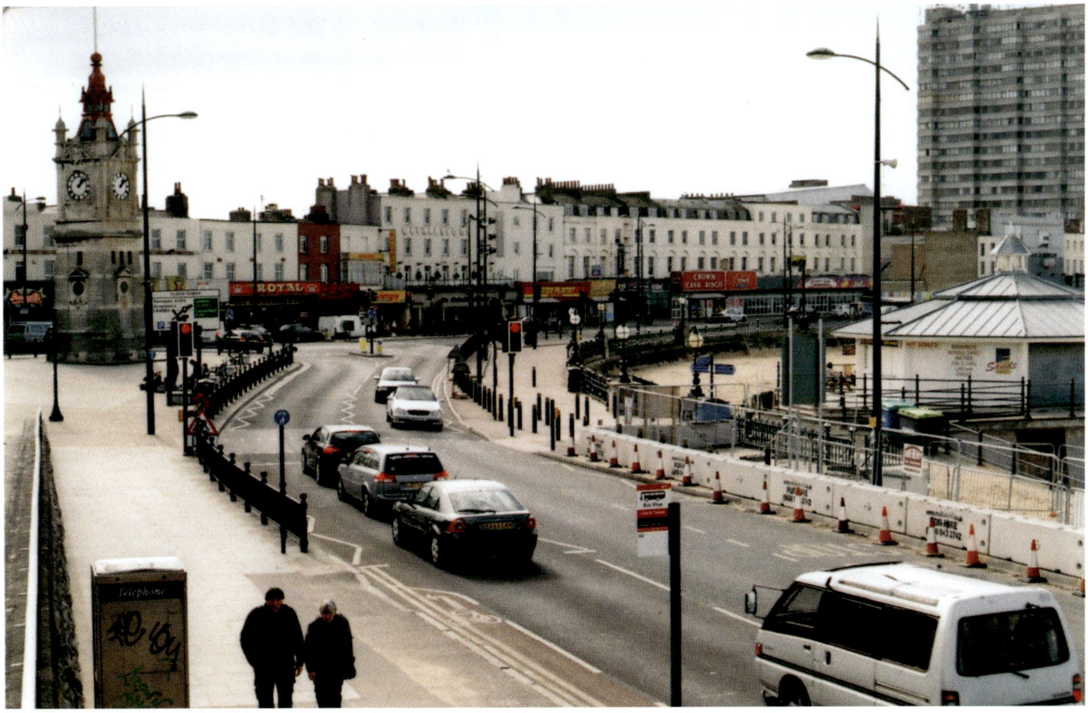

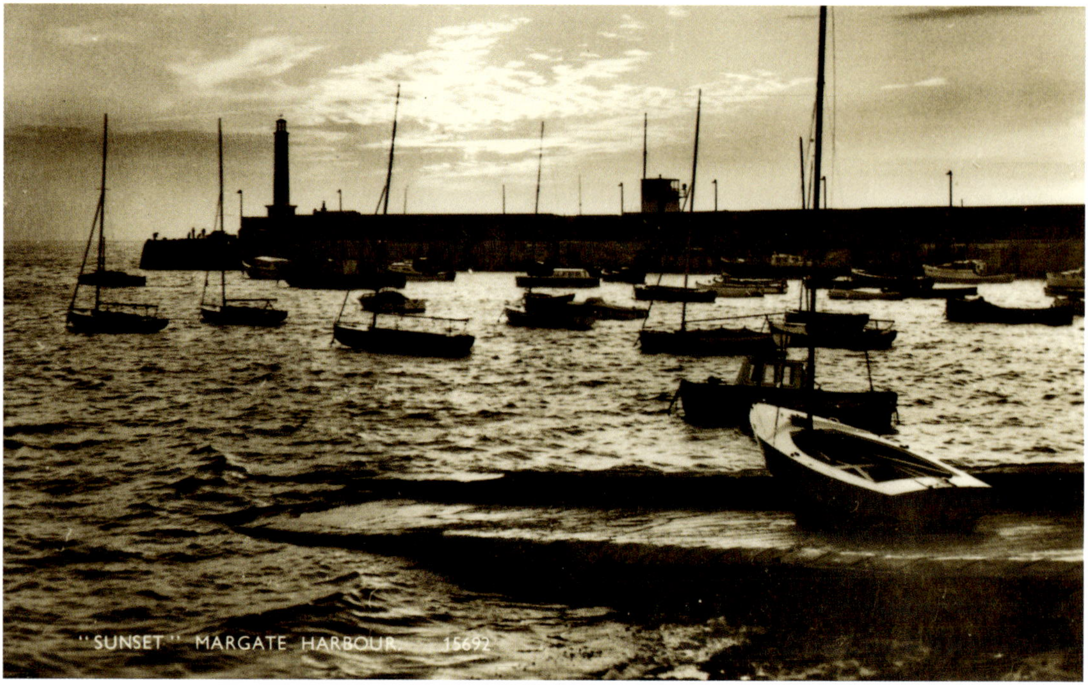

Margate Harbour

Margate's tiny harbour has never been a busy port, as these pictures attest. In later years it has become more of an anchorage for pleasure craft. It is hard to think now that once this was where the day-trippers and holidaymakers landed in their hoys and early steamships.

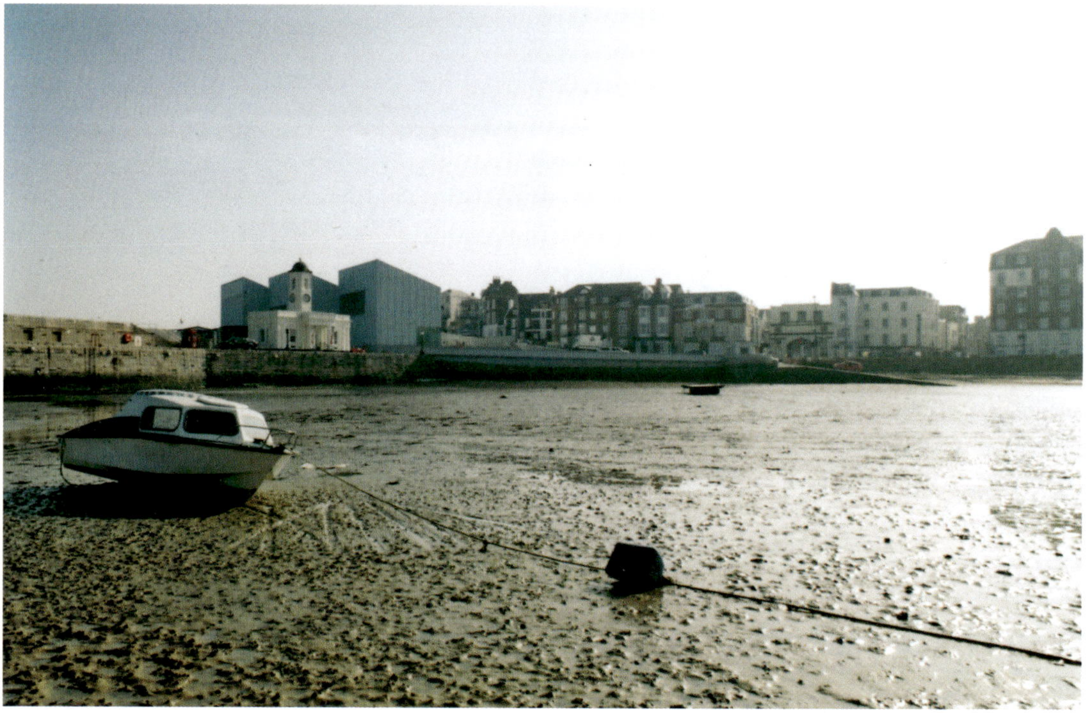

The Clock Tower

The Jubilee Clock Tower on the seafront below Marine
Gardens is an attractive piece of architecture. It was
built in 1887 to celebrate Queen Victoria's Golden
Jubilee, but was not completed until 1889 due to a gross
underestimation of the amount of work involved in
preparing the very substantial foundations required. The
local council appealed to the public for subscriptions
towards the project and set up a competition for the
design of the tower. Of the sixty-three entries received,
the design by Mr Kaufman was chosen. A total of
£2,100 was raised, but some of this money was used on
other events as well, so a cheaper design by Mr Henry
Arthur Cheers was used instead and the clock tower was
formally opened in a civic ceremony on 24 May 1889.
The tower holds five bells, above which was a time ball
which used to be raised just before 1 p.m. and dropped
on the hour; this has not worked for many years.
The bells were replaced in 1908 by Gillett & Johnson
of Croydon, but serious mechanical difficulties have
occurred over the years due to salt spray, meaning the
clock has often not functioned for long periods due to
the cost of repairs usually being quoted as quite high.
Being highly visible from the beach means that over the
years the clock tower has been most useful to those who
had trains or coaches to catch, or those with strict guest-
house mealtimes to meet. It was an important seafront
amenity because, when first built, wristwatches were yet
to be invented and few would have been able to afford a
pocket watch.

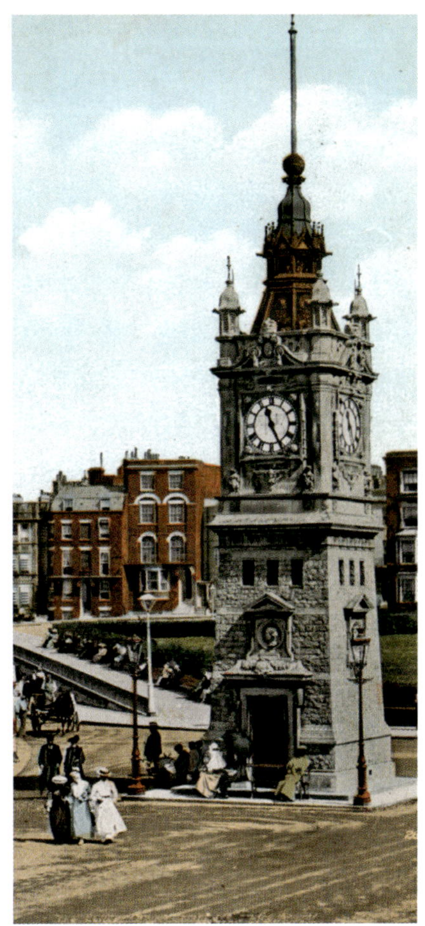

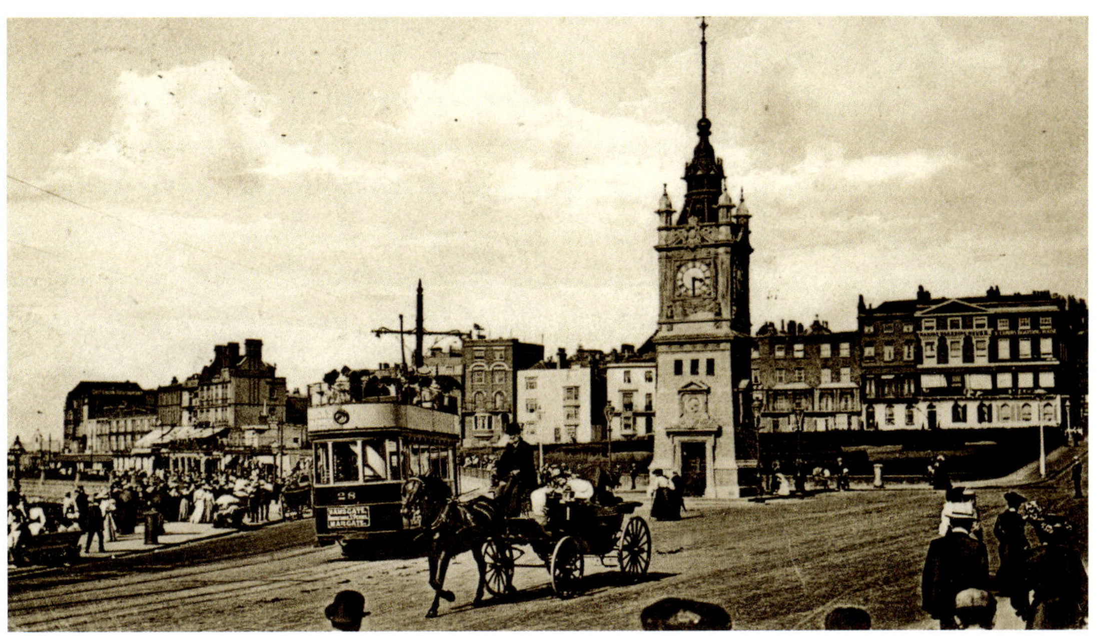

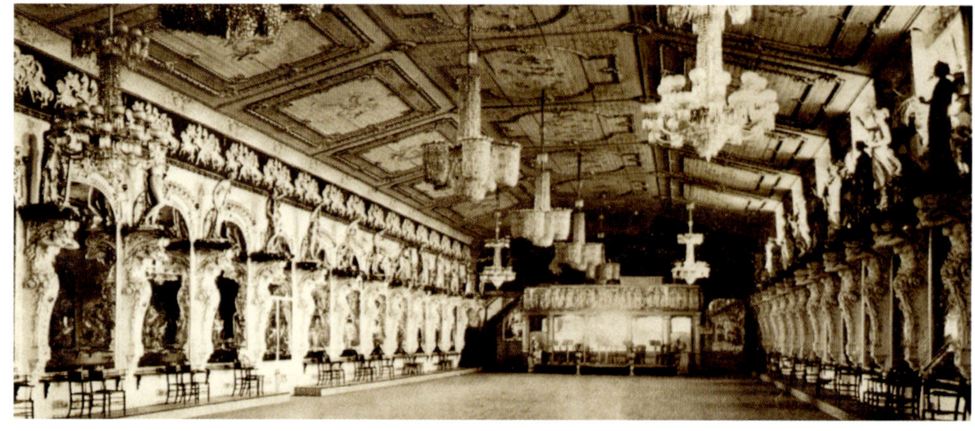

The Hall by the Sea

How many know that the origins of Dreamland lay in the arrival of the railway in town? In 1846 the South Eastern Railway completed a branch line from Ramsgate town station to a new railway station built on the present site of Arlington Arcade. Then in 1863 the SER's arch-rival, the London, Chatham and Dover Railway, built a terminus station on the site of what was to become Dreamland's cinema. However, upon extending its line to Ramsgate harbour station, the LCDR built a new through-station – that which is used to this day – to the west, making the terminus building redundant. It was bought by railway catering contractors Spiers and Pond, who opened a restaurant and dance hall in the unused railway terminus, naming it the Hall by the Sea. The venture was not a success and in 1870 was sold for £3,750 to the Reeve family of Margate, who also acquired additional low-lying land to the rear of the Hall. Later that same year, circus owner 'Lord' George Sanger went into partnership with Thomas Dalby Reeve, the then-mayor of Margate, to become the co-owner of the Hall by the Sea, and after Reeve's death in 1875 Sanger became the sole owner of both the Hall and the land immediately behind it. The land was turned into landscaped gardens with a 'ruined abbey', a lake, statues and a menagerie, together with sideshows and roundabouts. The zoo cages, which are unbelievably small, and Gothic-style walls on the west and south boundaries are now Listed Grade II structures and date from this era. One of the main purposes of the zoo was to act as a breeding and training centre for animals used in Sanger's travelling circus. This was how the Hall originally looked, but today it is lost amid years of structural changes. Only the approach road to the former amusement park serves to remind us in its name. Below we see the amusement park in its heydays of 1932.

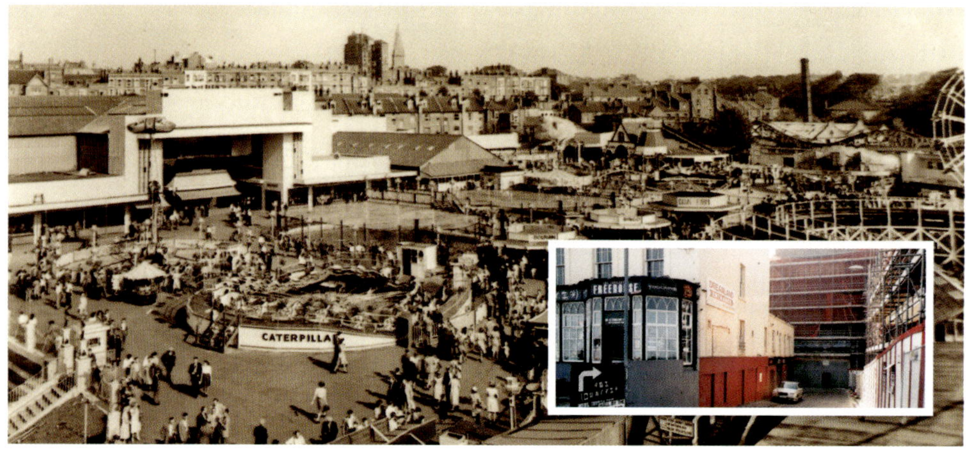

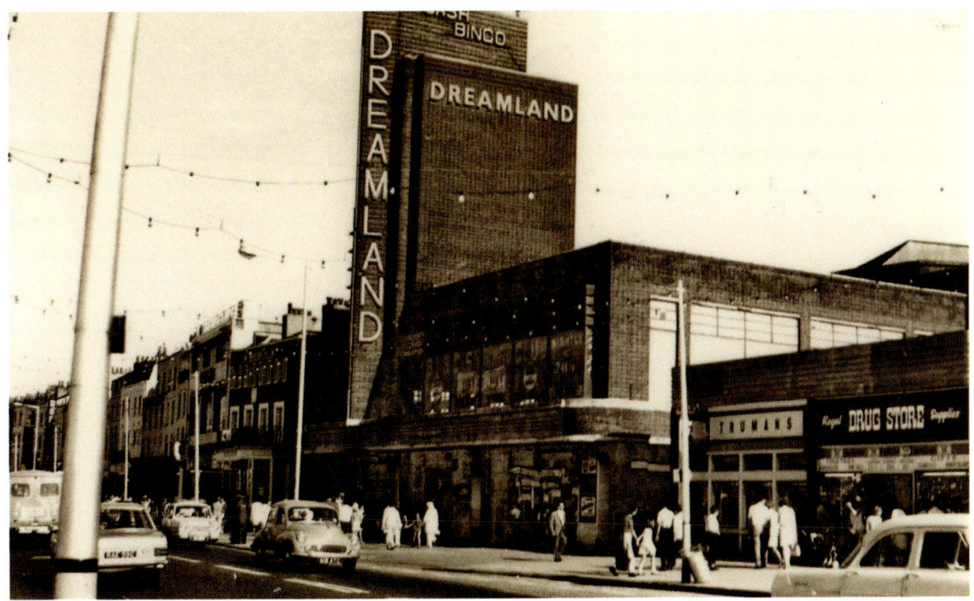

Dreamland Cinema

Dominating the seafront since it was built in 1935, this is Dreamland's former cinema. It was the earliest Art Deco cinema to be built in an expressionist brick idiom, and it heavily influenced the designs of the later Odeon cinemas. Its seating capacity was 2,050. It opened on 22 March 1935 with a show featuring *The Painted Veil* and closed in November 2007 with the showing of the 1957 film, *The Smallest Show On Earth*. The cinema closed for the war in 1940 and reopened on 1 July 1946. In 1973 the cinema was sub-divided and re-opened as two small theatres each seating 350, with a new bingo hall beneath. The cinema has had many owners over the years and was, until its closure in November 2007, operated by Reeltime Cinemas.

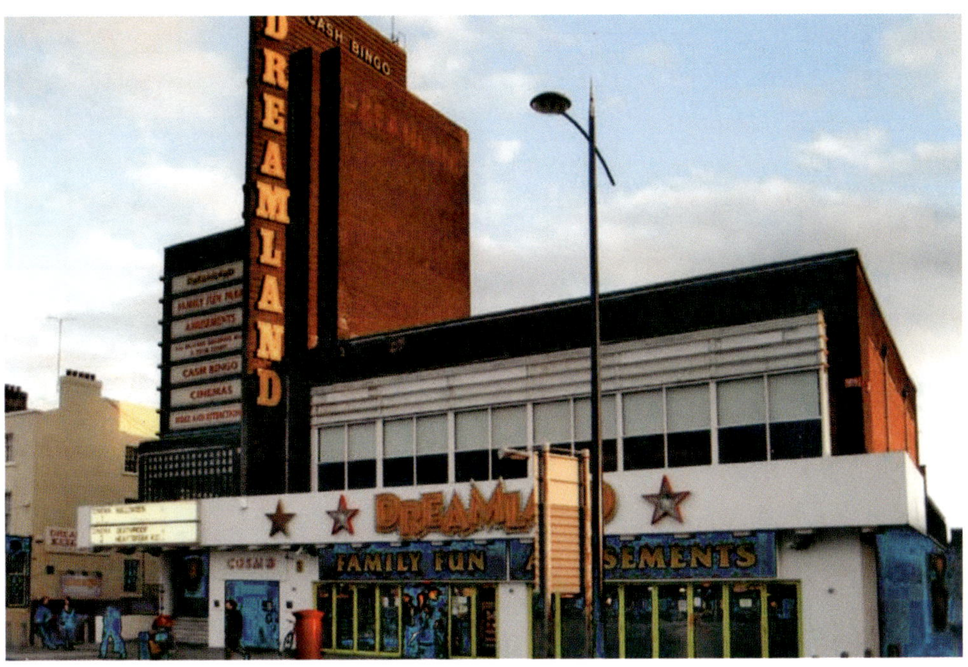

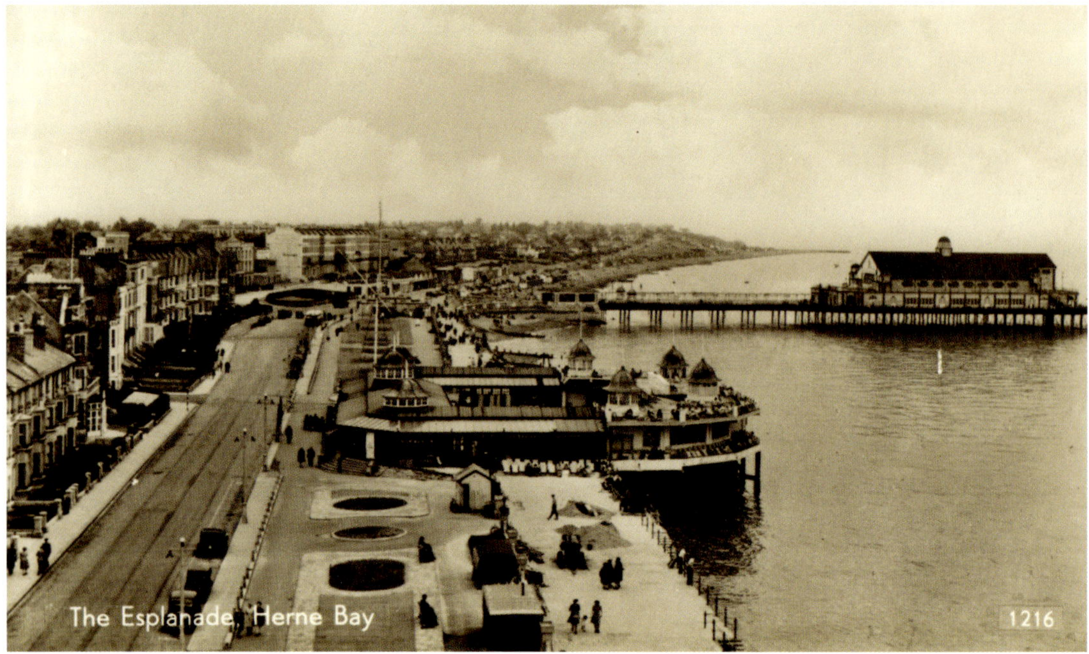

Herne Bay

Another resort that was within easy reach of the early Victorian steamships was Herne Bay. At the beginning of the nineteenth century Herne Bay had a population of fewer than a dozen inhabitants, but a nearby military encampment prompted an expansion of the population. This in turn attracted visitors who disembarked from the passing steamships, but after a few bumpy rides in the hoys that were used to bring them ashore, the visitors decided they needed a pier and family accommodation at Herne Bay, and so work on the first Herne Bay Pier began. The railway station was built in 1861, thus beginning the town's growth into a holiday resort. Both of these views were taken from the top of the clock tower.

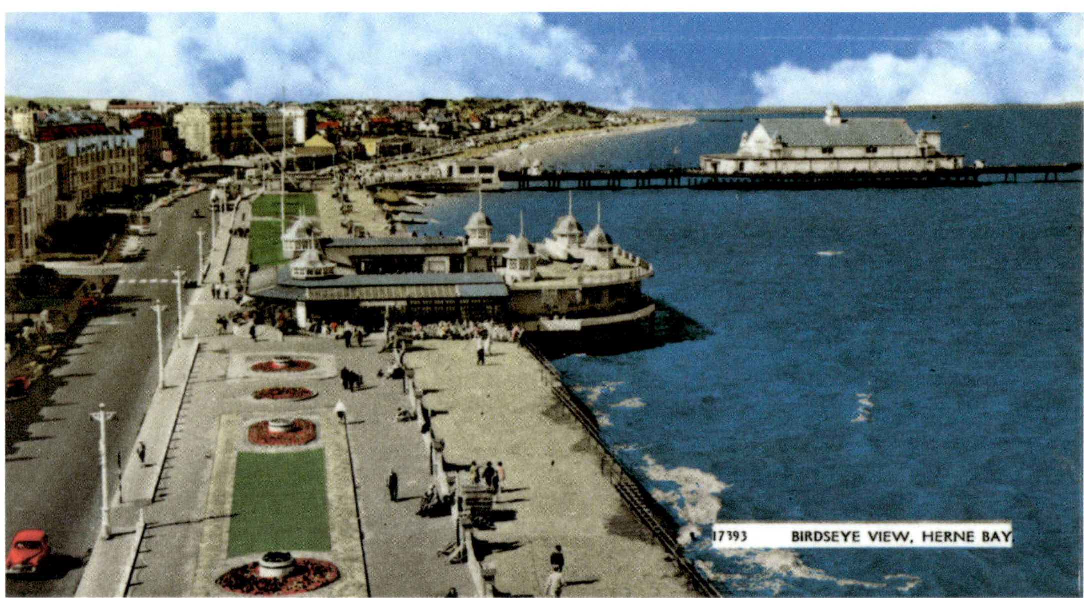

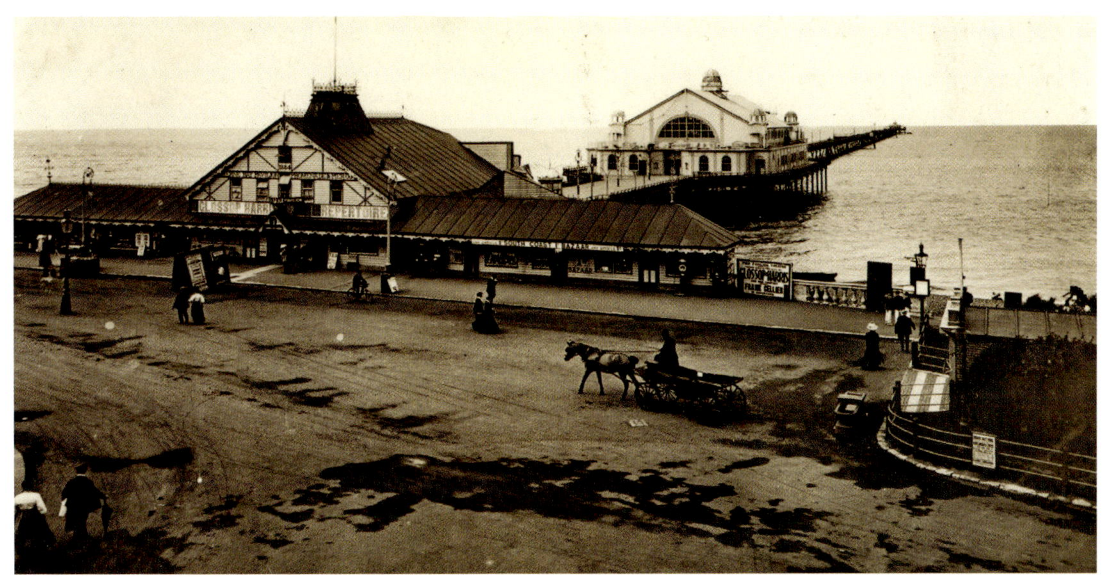

The Story of the Pier

The pier you see today is Herne Bay's third pier. Its stunted appearance can be explained by the fact that the pier was destroyed in a storm in 1978 and dismantled in 1980, leaving just a short stub with a sports centre at the landward end; the pier-head still stands, isolated, out to sea. It was preceded by two other piers, a wooden deep-sea pier designed by Thomas Rhodes, opened in May 1832 at a cost of £50,000, and a second, shorter, iron version by Wilkinson & Smith that opened in August 1873, costing only £2,000. A feature of the first pier that remained right up to 1953 was a curved stone balustrade at the entrance, taken from the old London Bridge which was demolished in 1831. Such was the damage caused to the balustrade by the notorious East Coast floods that it had to be replaced by iron railings. The whereabouts of the balustrade today remains a mystery.

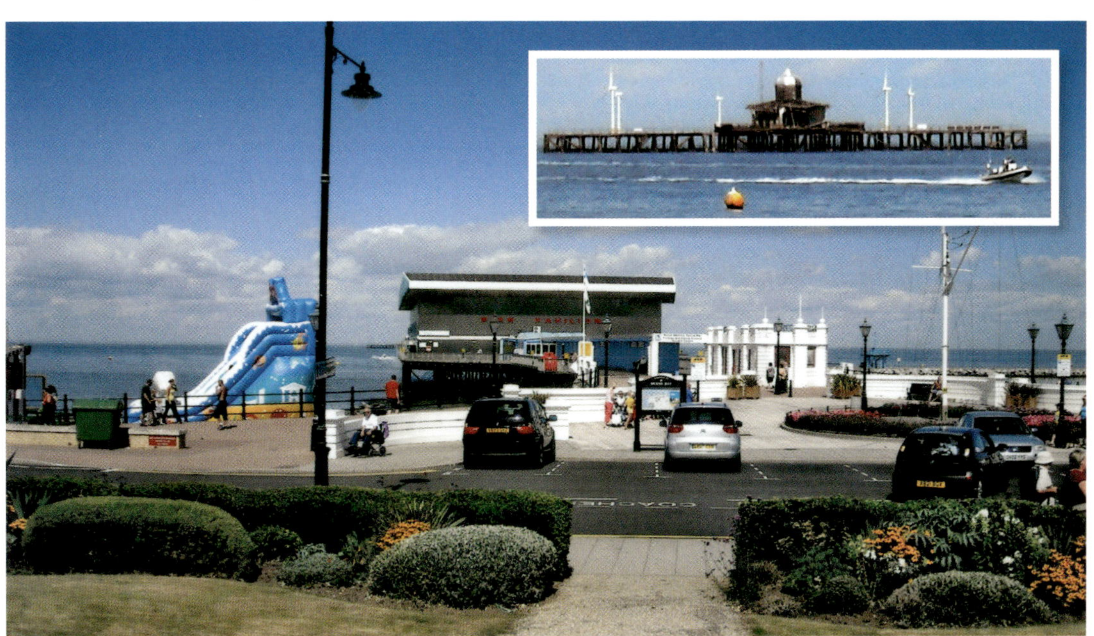

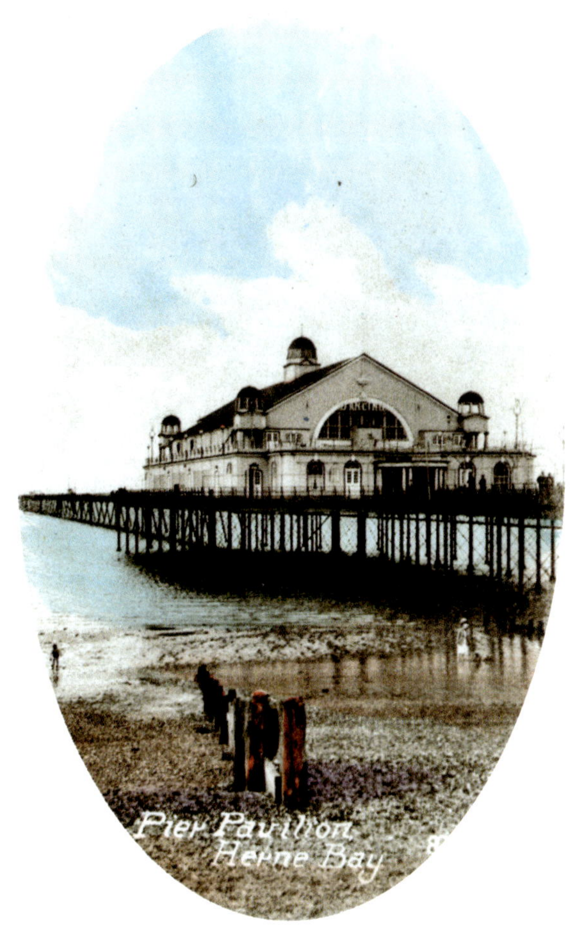

The Pier Pavilion

The Herne Bay Pavilion Pier and Promenade Company built a wooden theatre, shops, lavatories and a ticket office across its entrance in June 1884. The theatre, known as the Pavilion, was designed by McIntyre North and opened on 24 July 1884. In 1895 the second pier was extended to give the pier we have today. In building the third pier, the theatre was retained but in October 1908 the Receiver offered the pier to Herne Bay Urban District Council for £6,000. The purchase was completed in September 1909. The council organised a competition to design a new Grand Pier Pavilion, and in May and June of that year the marquee section was widened and the Pavilion built for £2,000. The Grand Pavilion survived a fire on 9 September 1928 when the theatre, shops and Mazzoleni's Café at the entrance were all destroyed. In 1932, the Pier Approach was redeveloped to replace the fire-damaged site. Here we see two contrasting views of the Pier Pavilion, the top one taken in the early 1900s, while the bottom view is how it looked later.

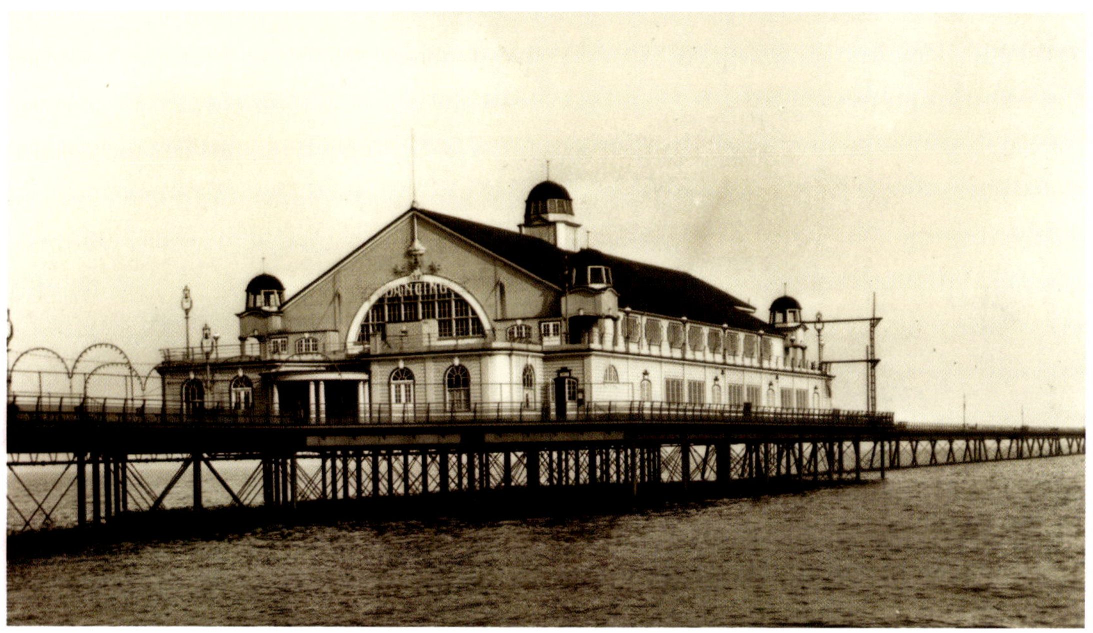

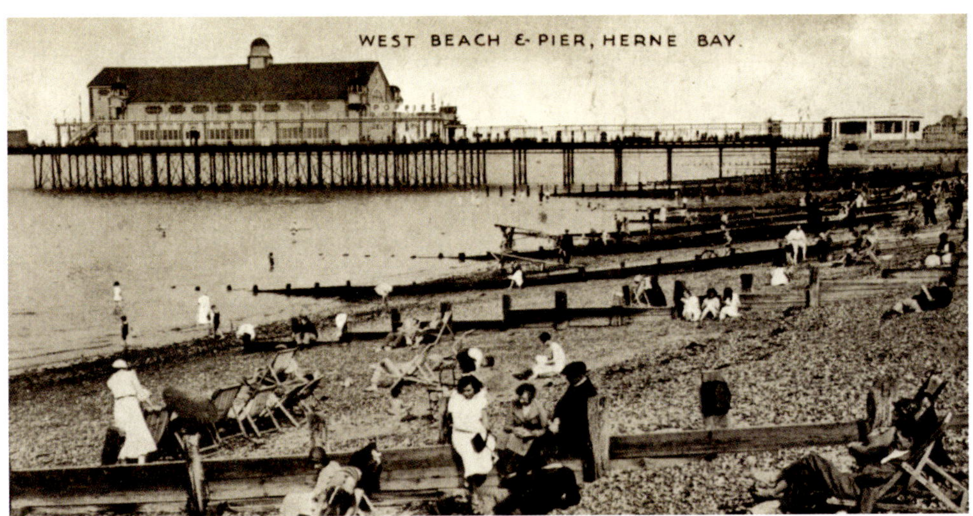

The Pier Pavilion in Later Years

By 1968 the seaward end of the pier had been closed and abandoned, and insurance withdrawn for the part of the pier beyond the pavilion; the public were excluded. The Grand Pier Pavilion was refurbished at a cost of £158,000 but was destroyed by fire, possibly caused by a spark from a welding torch during pier-entrance reconstruction in June 1970. The building burnt down within hours, much to the distress of the residents of Herne Bay. As a replacement, the Pier Pavilion was designed in 1971 by John C. Clague, opened by the Rt Hon. Sir Edward Heath on 5 September 1976, and disparagingly called 'The Cowshed' by the public. It certainly lacks any architectural merit. In the spring of 2009 Canterbury City Council agreed to the formation of the Herne Bay Pier Trust whose main objective is the preservation, renovation, reconstruction and enhancement of the pier. Herne Bay Pier was also unusual in that the Pavilion housed the Pier Sports Centre, which offered gym, dance and sports facilities and boasted a roller skating rink which, besides providing facilities for individuals and families, was the home of some of the best roller hockey teams in the country. The Pavilion was condemned in September 2010, and by March 2012 demolition work was well underway. What will replace it remains to be seen.

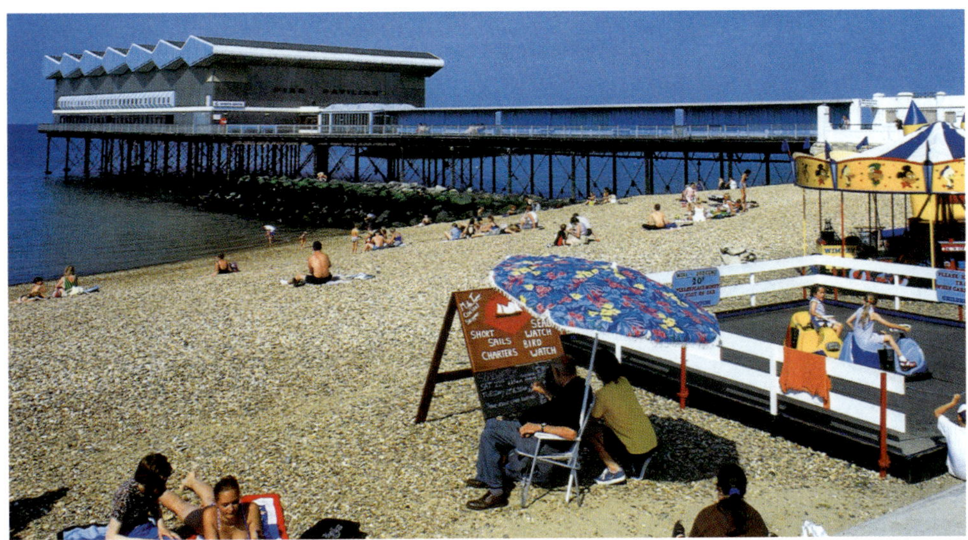

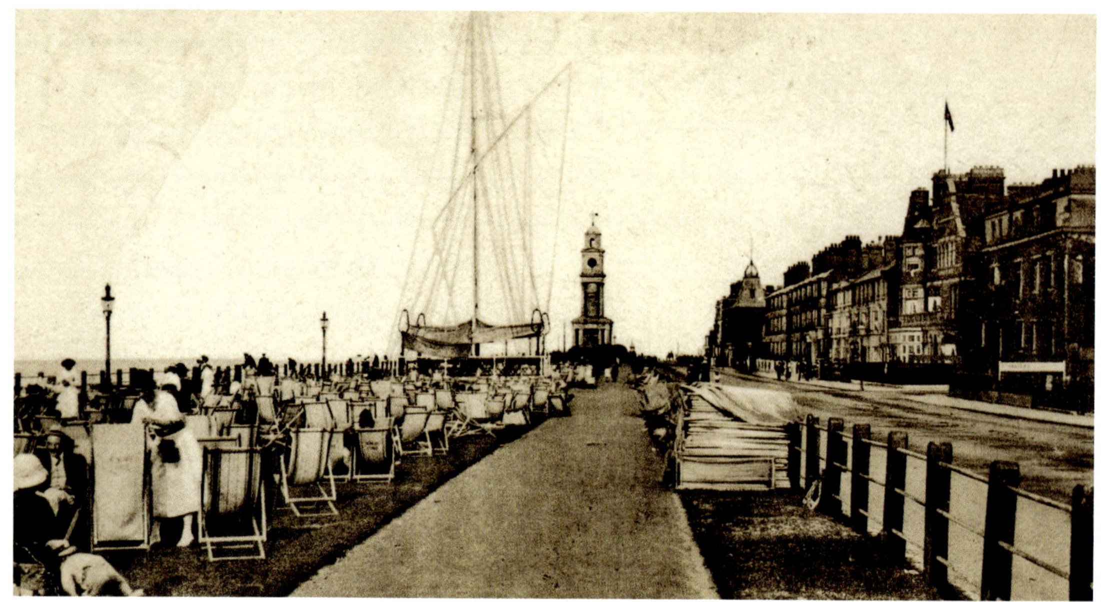

The Tower Gardens

Even though Herne Bay had acquired a pier and become popular with day-trippers by 1832, its seafront did not come into being until much later. The present town was founded in the early 1900s by London speculators who, noting the unspoilt charm of the bay, planned a new resort to be called St Augustine's. The name did not catch on and it continued to be known as Herne Bay after the nearby village, Herne. It was during this period that a wealthy London lady gave the town its distinctive 80-foot clock tower, which dominates the promenade. What had once been the haunt of smugglers and fishermen had grown into a fashionable Victorian resort with all the attendant features of bathing machines and assembly rooms. It was redesigned in 1990 as part of the strengthening of the sea defences to include exciting, modern sculptures among the floral displays and a children's play area.

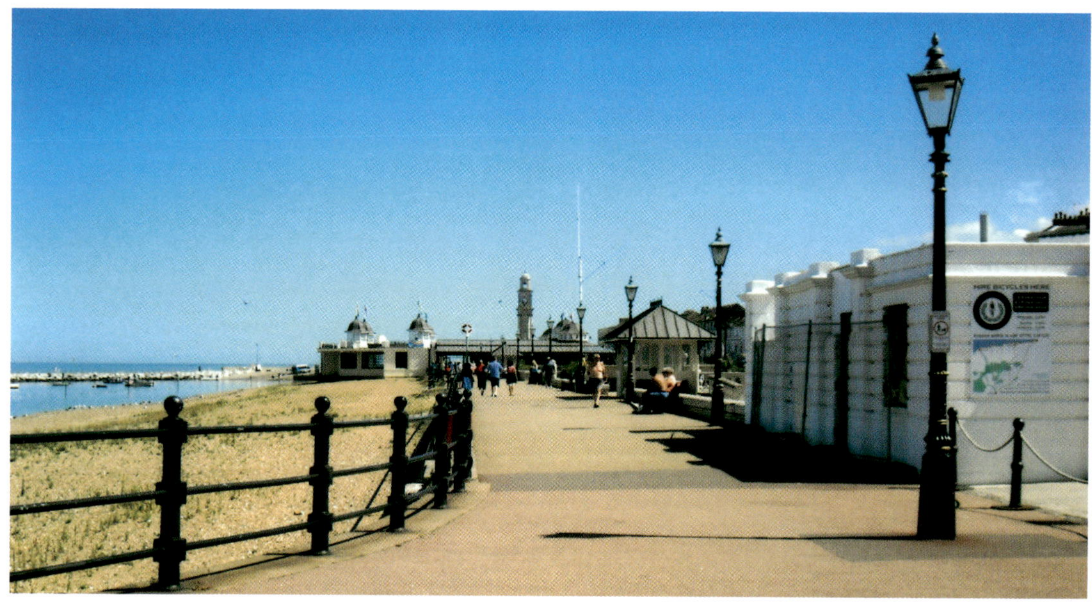

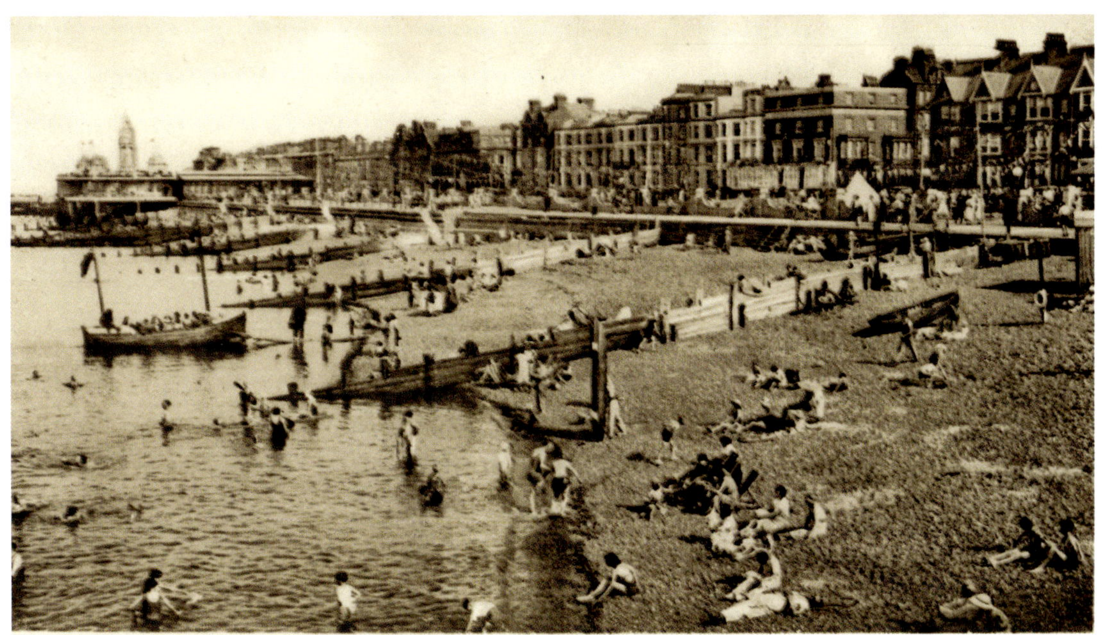

A Changing Beachscape

How very different the beach looks from the pier in these two views; the top one is probably taken during the 1930s but posted at the end of the war in 1944, and the 2012 view is below. It appears that the sun-seekers have been replaced by grass! In truth this beach is not now as popular as that to the west of the pier, as a mini-harbour has been built here, just off-shore, where weekend sailors can launch their vessels before putting out to sea. The postmark on the top postcard is 25 September 1944; D-Day and the invasion of Europe had taken place only three months earlier and it was yet to be known whether the Allies had succeeded or not. The beach at Herne Bay, like all other coastal resorts, would still have been littered with anti-invasion measures, including sections of the pier missing. This postcard's message does not make a single mention of this, only a view and a message of someone enjoying their holiday.

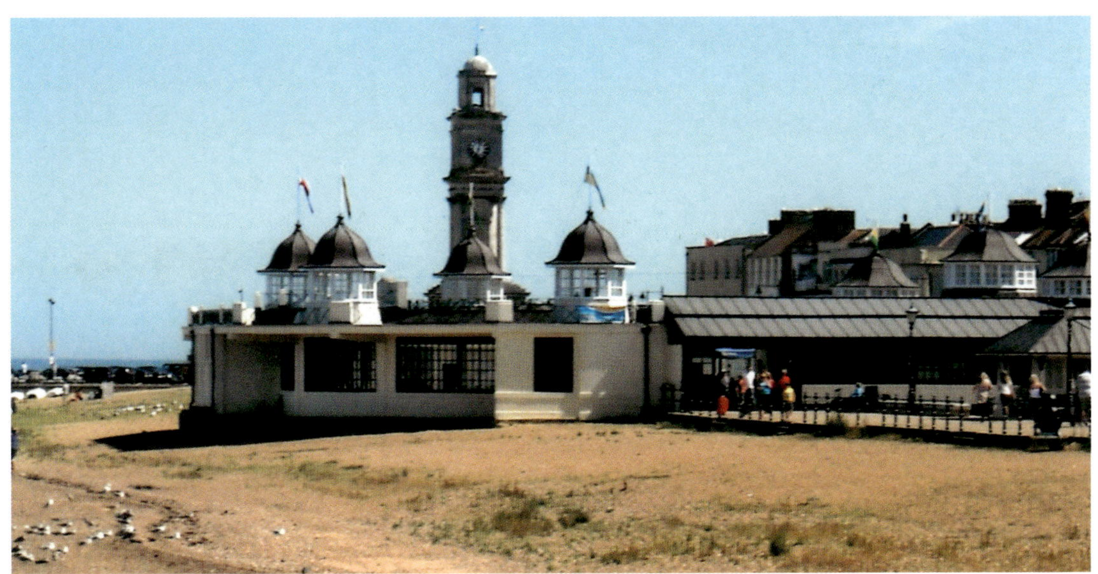

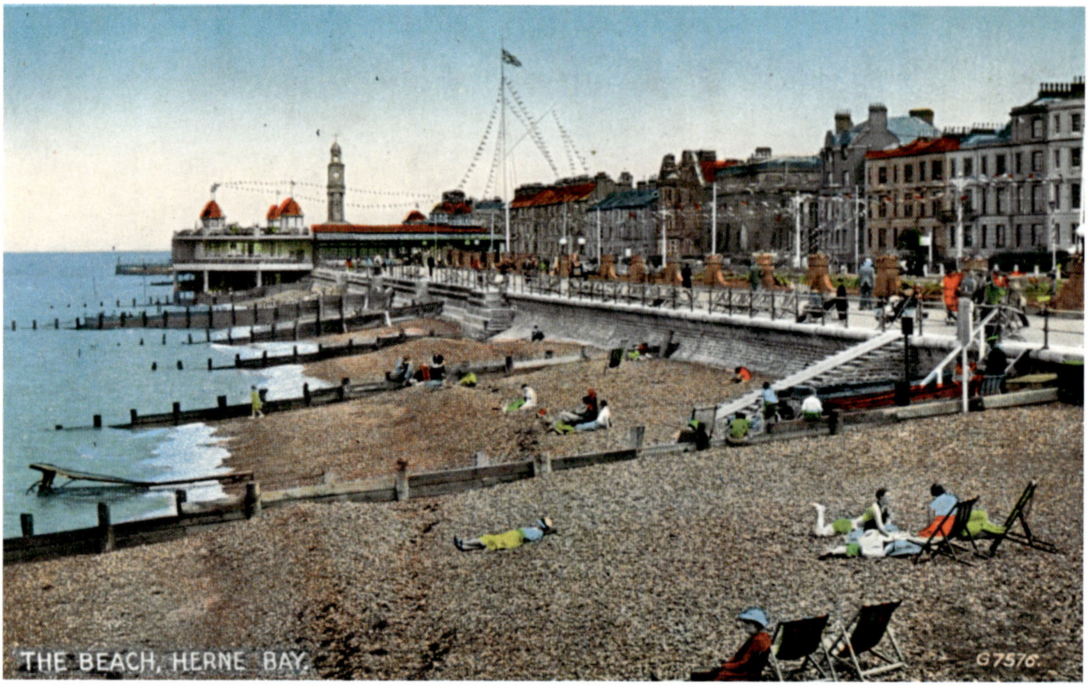

The Bandstand from the Pier

An undated, hand-tinted view looking towards the bandstand, where little seems to have changed in a century or more except the number of people now using the beach. The mini-harbour referred to above can be seen to the left of today's view.

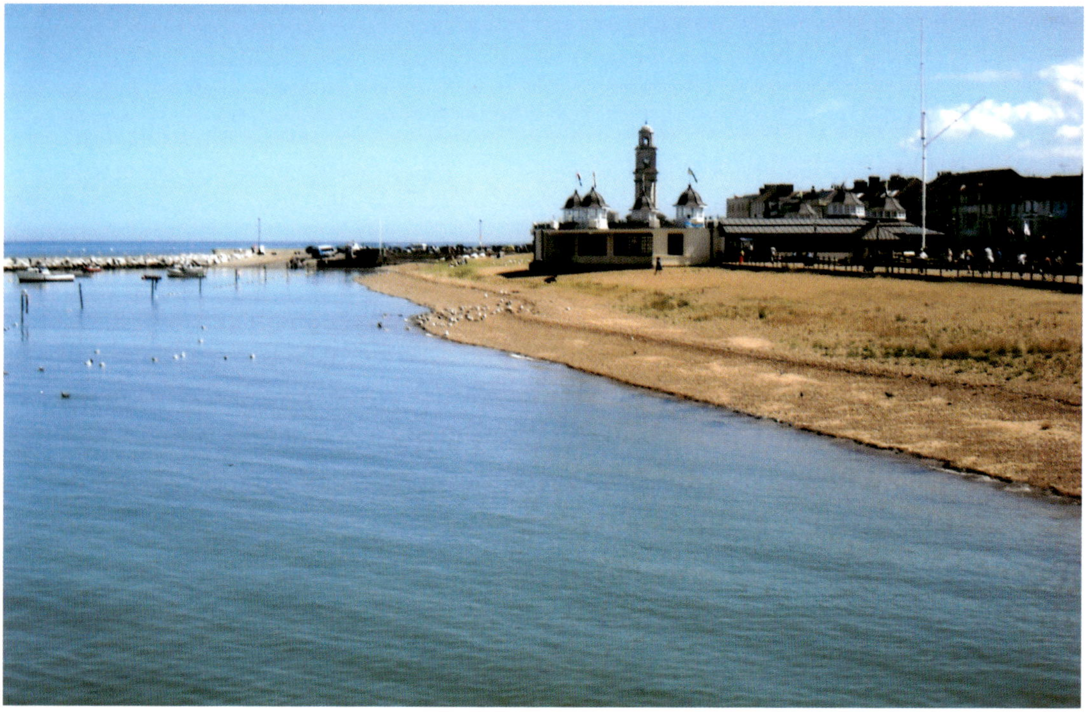

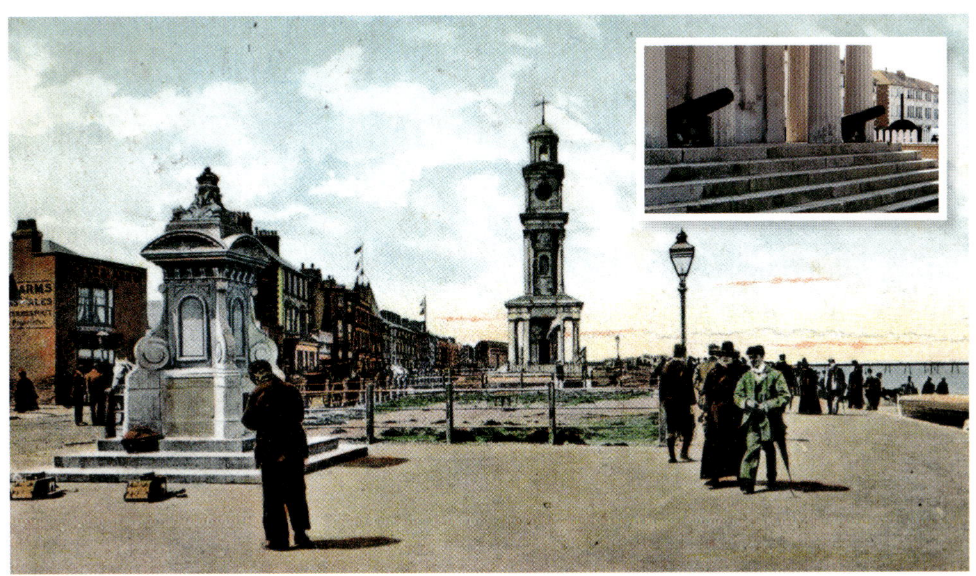

The Clock Tower

Herne Bay's clock tower is unquestionably the most striking piece of architecture in the town. It was the gift of a widow and benefactor to the town, Mrs Ann Thwaytes, who when visiting Herne Bay lived at No. 30 Marine Terrace. She commissioned architect Edwin Dangerfield to design it and it is said he based his design on a building depicted on the Guardian Royal Exchange Insurance fire badge affixed to Mrs Thwaytes' house. The foundation stone was laid on 3 October 1836 and the tower, with a brick interior and Portland stone cladding, was completed by 5 October 1837. It stands 82 feet high to the weathervane, and cost £4,000. On its northern face is a memorial plaque to commemorate the men of Herne Bay who volunteered for the Boer War. On each side of the plaque on the walkway are two cannons on wooden trucks, which were dredged from the seabed when the town's third pier was being constructed; they are thought to be of Dutch origin. When first found they were used for firing blank shot as a fog warning on the pier. Many seaside resorts throughout Britain have a seafront clock, which was an important seafront amenity because when first built wristwatches were yet to be invented and few would have been able to afford a pocket watch.

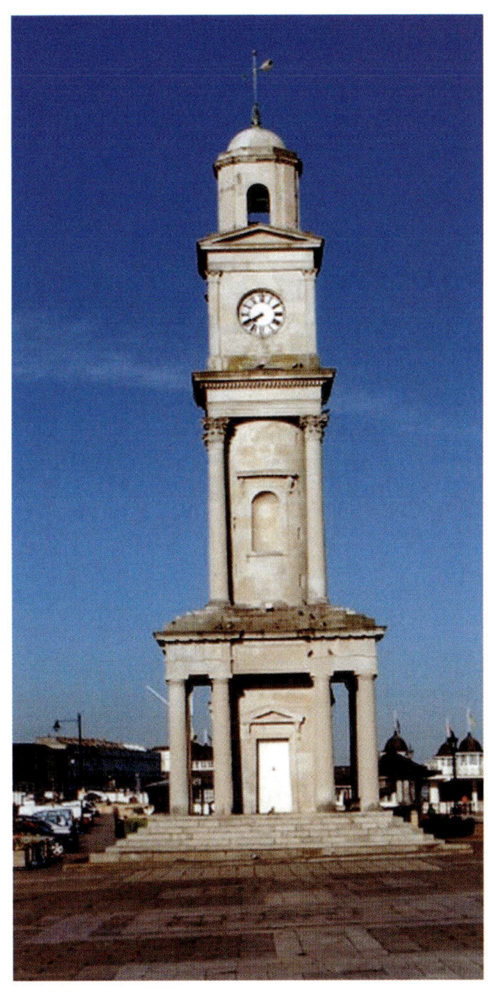

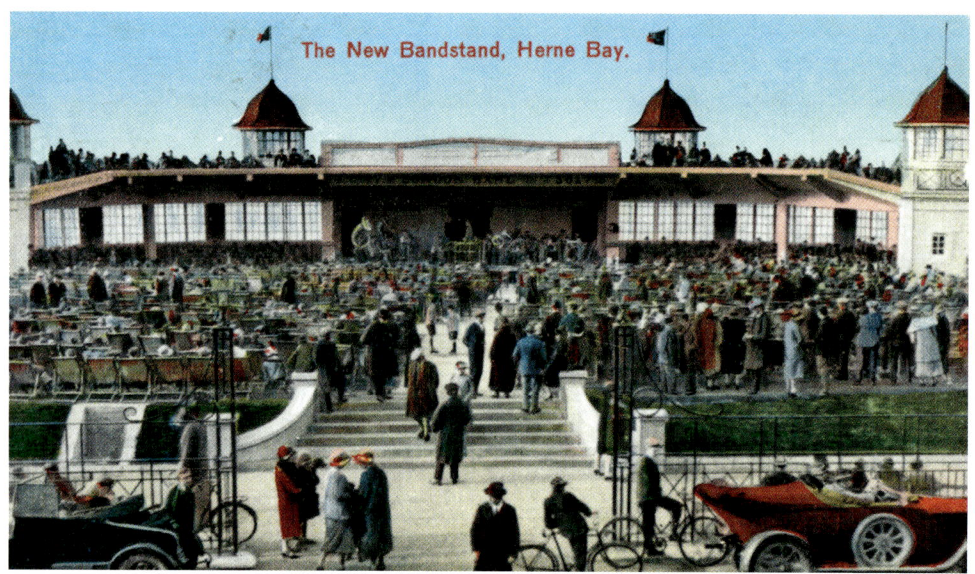

The Bandstand

An essential seaside amenity for the Victorian and Edwardian holidaymakers was the bandstand, where they could be enjoy not only their favourite brass bands but many other forms of entertainment as well. Like the clock tower, it became an essential amenity in most resorts. Herne Bay's bandstand was built in 1924 along with the rest of the promenade, which until then had been just a grassy strip of coastline. To begin with it was a crescent-shaped, three-sided building whose south face was open to the elements. Screens were available to protect the public from the wind blowing in from the east or the west. The roof over the stage area was used as a sun deck, while below there was a promenade deck overlooking the sea. Herne Bay is believed to be one of the first seaside resorts in the south of England to offer first-class military bands playing to packed audiences daily throughout the summer. During performances, passing motorists were asked not to hoot their horns during performances. The top picture, taken in 1931, just three years before the enclosure work began, shows how popular these venues were.

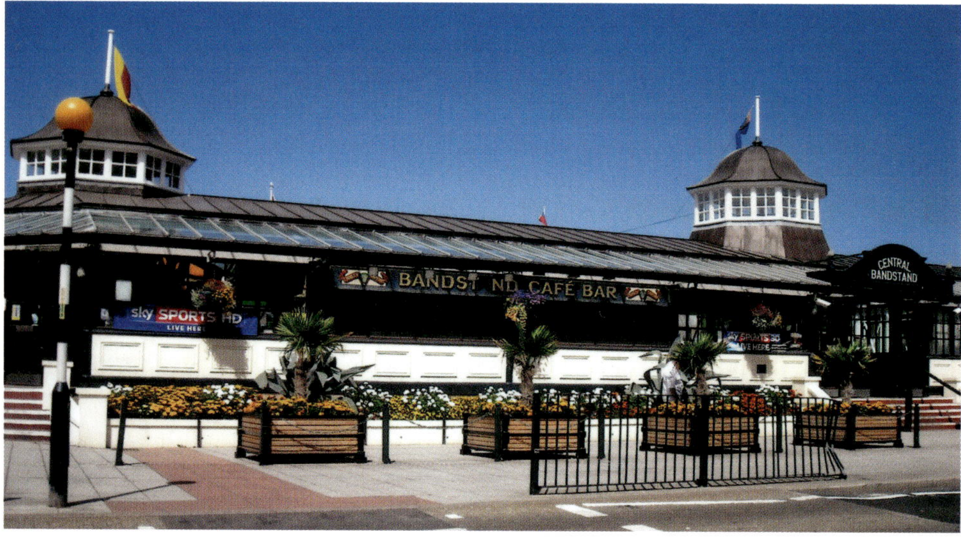

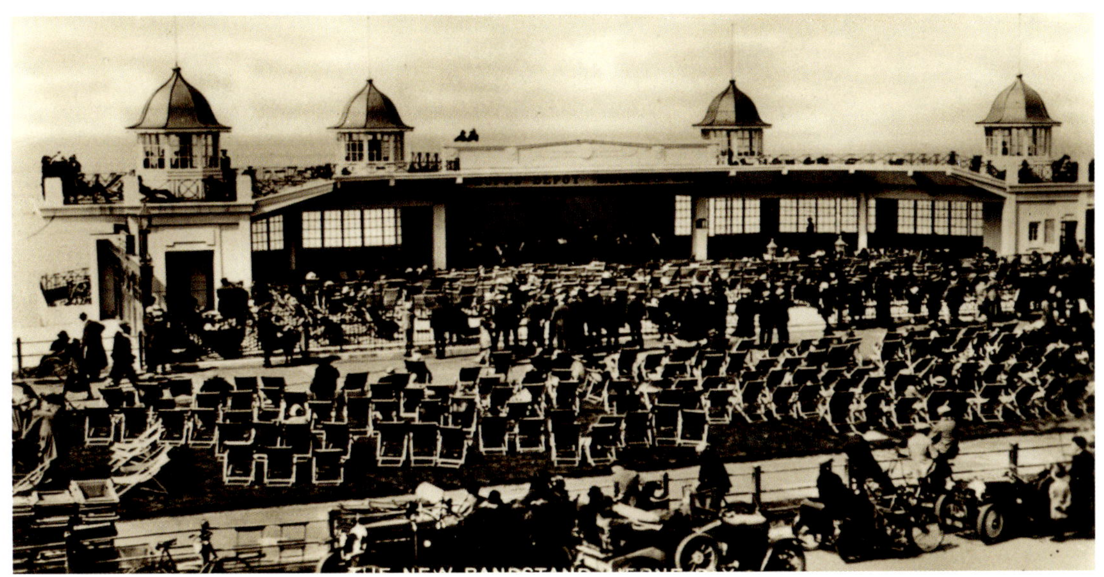

The Bandstand in Later Years

In 1934, the front section abutting on to the road, which today is used as a café, was added, thus enclosing the entire building. Glazed roofing panels were added to the side sections, but the centre was left open to the sky, just as it still is. It became a popular venue for afternoon tea dances with live bands. There were also Punch and Judy shows and magic shows held here for the children. The bandstand continued to be popular right up to the outbreak of the Second World War, when all seafronts became a 'no-go' area; beaches had to be defended against possible invasion. But soon after the cessation of hostilities, the bandstand quickly regained its pre-war popularity and continued to entertain holidaymakers and the locals alike until the 1980s, when it started to show its age and fell into a state of disrepair. It was restored to its former glory in 1999, when the copper domes and glazed ironworks were added. It is such a shame that today's modern street-furniture-like railings and pedestrian crossings mar such a fine old building.

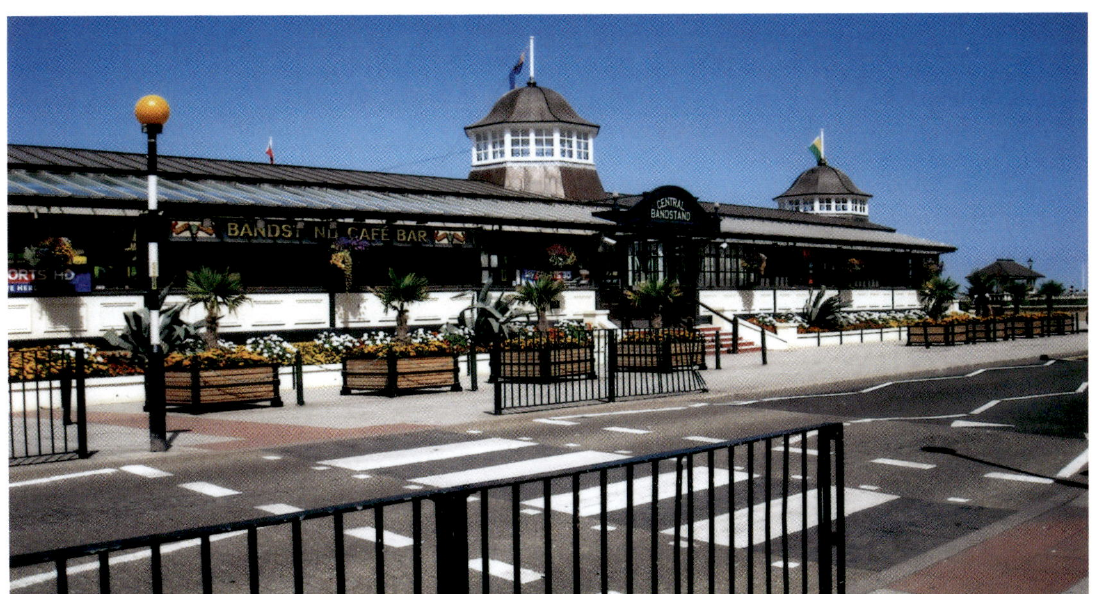

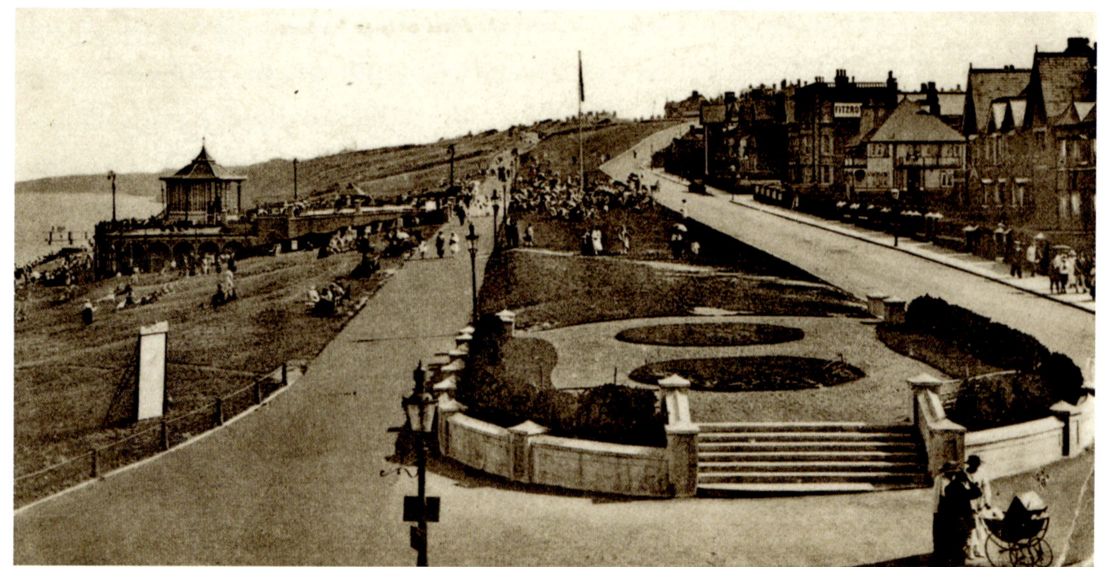

The Downs and East Cliff

Should visitors to Herne Bay tire of the beach or promenade, then a brisk walk could be taken along the Downs, a grassy area on the East Cliff, towards Reculver with its ruined church, which is slowly slipping into the sea. Reculver was also once the site of a Roman fort that guarded the mouth of the Wansum Channel, a stretch of water that separated the Isle of Thanet from mainland Kent. *En route* they could stop awhile at the small bandstand seen in the top picture, which was built in 1904 and was known simply as the Pavilion or the Bandstand. This later became the King's Hall, which was originally named the King Edward VII Memorial Hall, and was opened on 10 July 1913 by HRH Princess Henry of Battenberg, on behalf of Queen Alexandra. Part of the transformation included rebuilding and extending it underground.

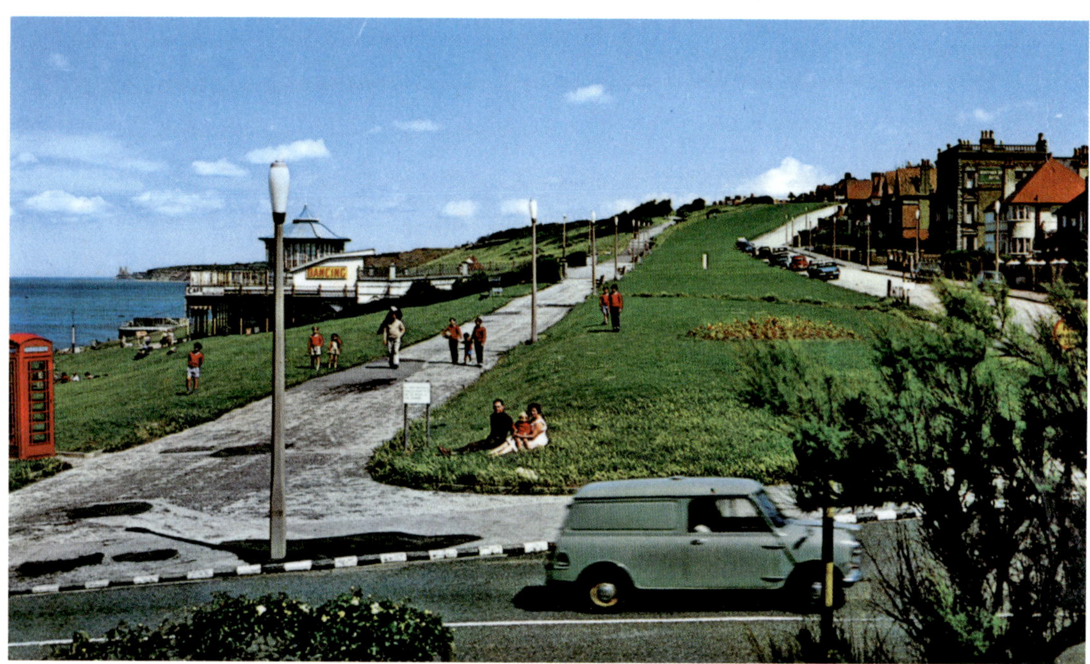

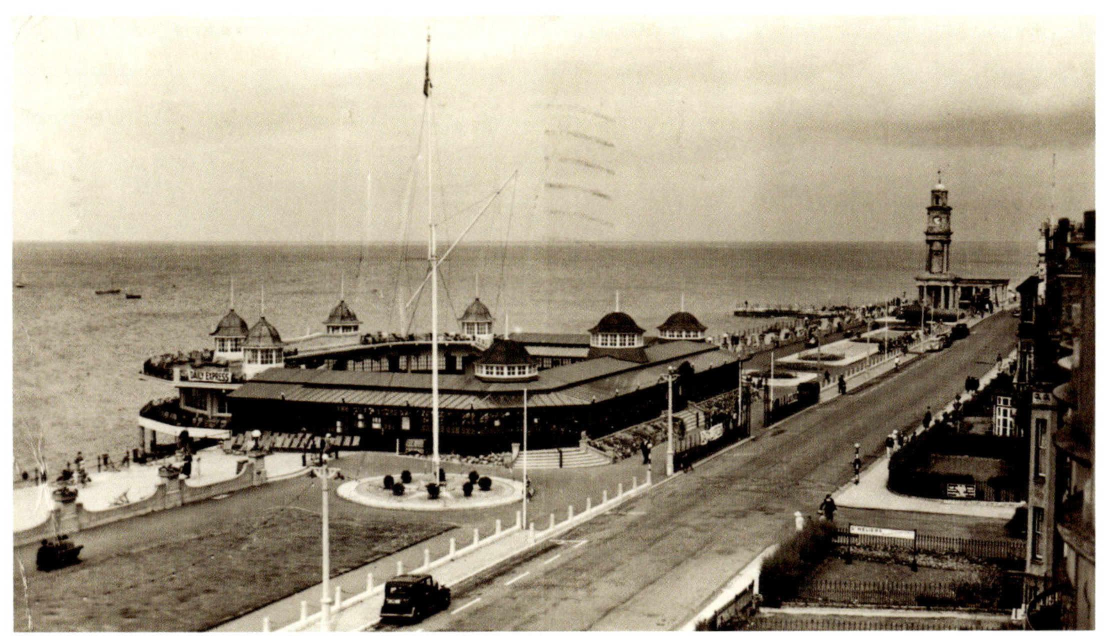

A Final Look at Herne Bay

Little seems to have changed in Herne Bay. The top picture, taken in 1939, is remarkably similar to the lower picture, taken around the 1960s/70s. After reading the message on the reverse of the top postcard, I wonder if either the sender or recipient knew that the Second World War was only one month away. The summer of 1939 was to be the last holiday for most for many a year. It is only within the past ten to fifteen years that noticeable differences to the seafront have been made. Extensive seafront regeneration took place in the 1990s, including the restoration of the Victorian gardens, which are now a delight to sit in, and the bandstand was refurbished and brought back into use for the public.

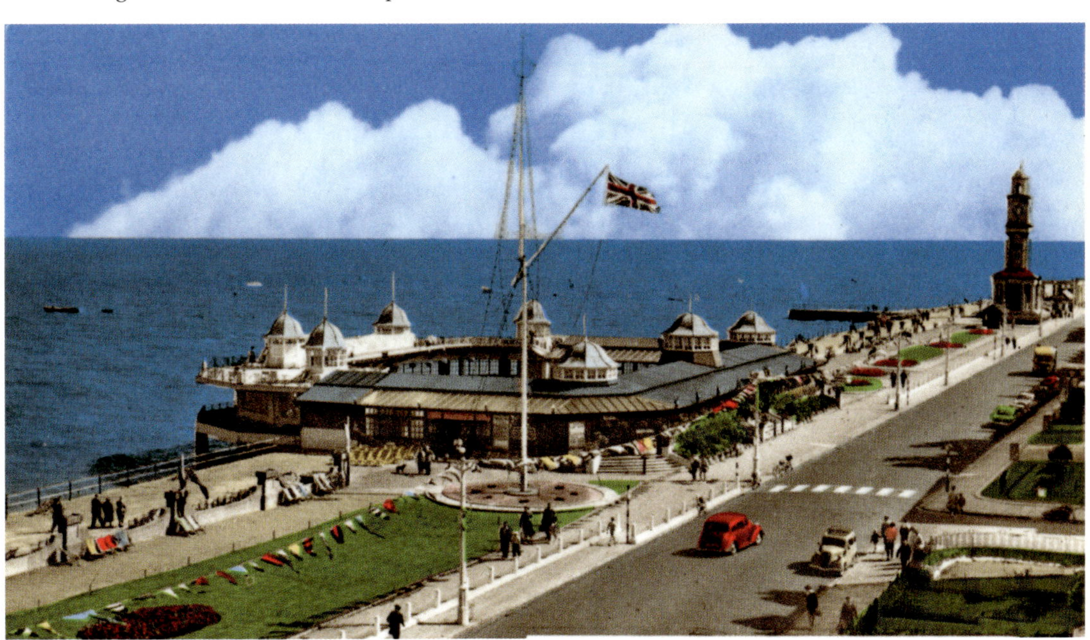

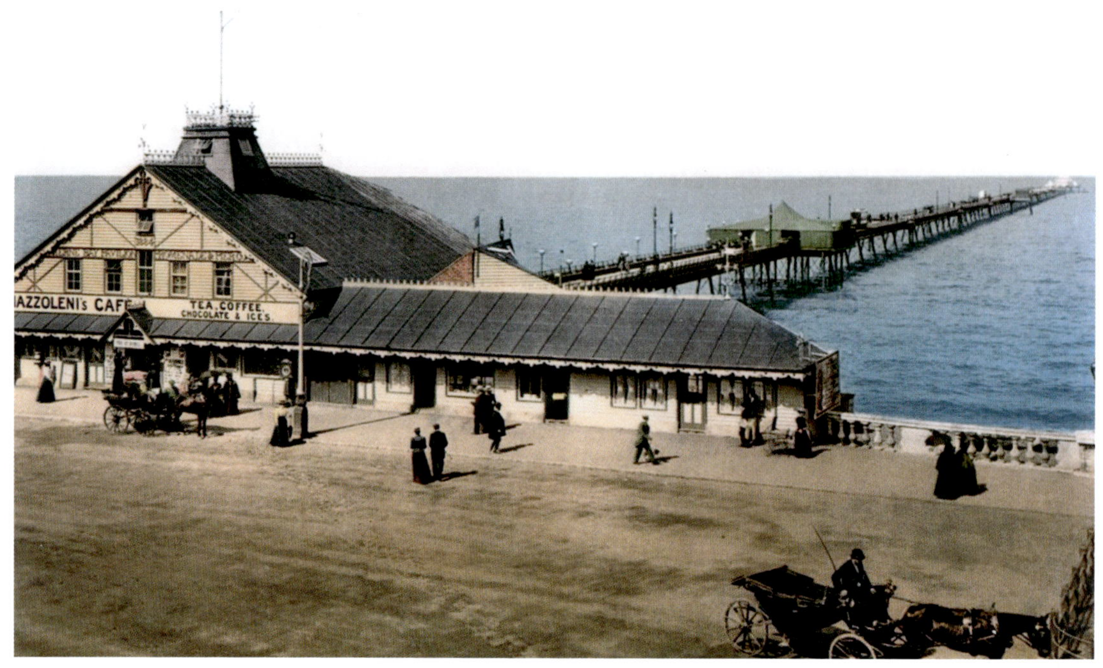

What Will Become of the Pier?

By way of a fitting end to the story of Herne Bay's development as one of Kent's leading seaside resorts, I can do no better than to end where we began, with the pier. The top view is of the pier, taken around 1900. On it can clearly be seen the curved stone balustrade taken from the old London Bridge, the marquee area that was widened when the Grand Pier Pavilion was built in 1910, and the theatre, shops and Mazzoleni's Café at the entrance. How very different today's pier looks. A severe storm in 1978 reduced its length to almost nothing and in the spring of 2012 the sports hall was demolished. Where it once stood is today a tarmac-covered area where different activities will be held.

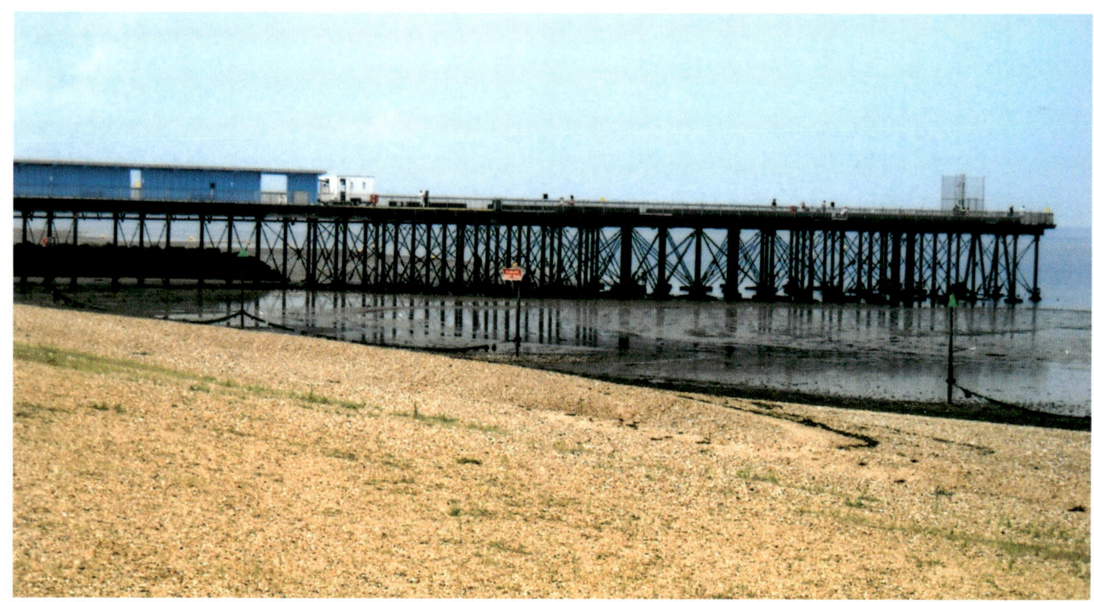

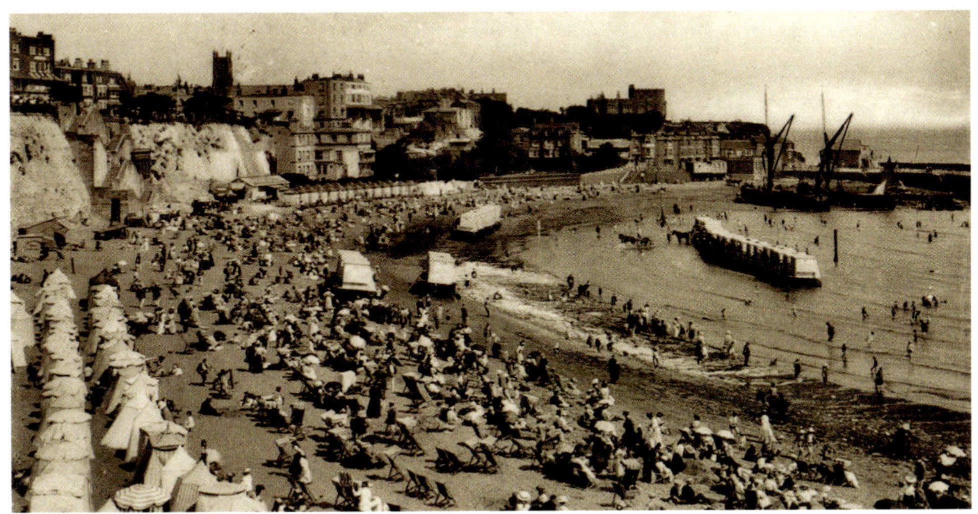

Broadstairs

Just a few miles east of Margate is the fishing harbour of Broadstairs. By 1824 steamships were becoming a more common form of transport, having taken over from the hoys and sailing packets in around 1814. They made trade with London much faster. Mixed feelings must have been strongly expressed by the Thanet boatmen as the unrivalled speed of the new steamships outmanoeuvred all other classes of vessel, but it brought a new prosperity to Thanet and by the middle of the nineteenth century, the professional classes had begun to move in. Steady town expansion took place, the population doubled in fifty years to 3,000, and as the town grew, artists, writers and poets started to visit, as did clerks, lawyers and architects, causing more accommodation for rent to be built. Broadstairs' seaside holiday industry had started. Although many holidaymakers were attracted to Broadstairs during the Victorian era, the town was not directly served by the railways until 1863, a time of great expansion for railways in the south-east of England. Rail access to Broadstairs had previously relied heavily upon coach links from other railway stations in the district. Broadstairs' railway station is a ten-minute walk from the beach, and was rebuilt in the 1920s. This view of August 1914 shows how popular Broadstairs beach was, with bathing machines jostling for foreshore space with the sailing ships moored in the harbour. It is still as popular today, but to be fair, these modern-day views of Broadstairs beach were taken early in the year before the summer season had begun.

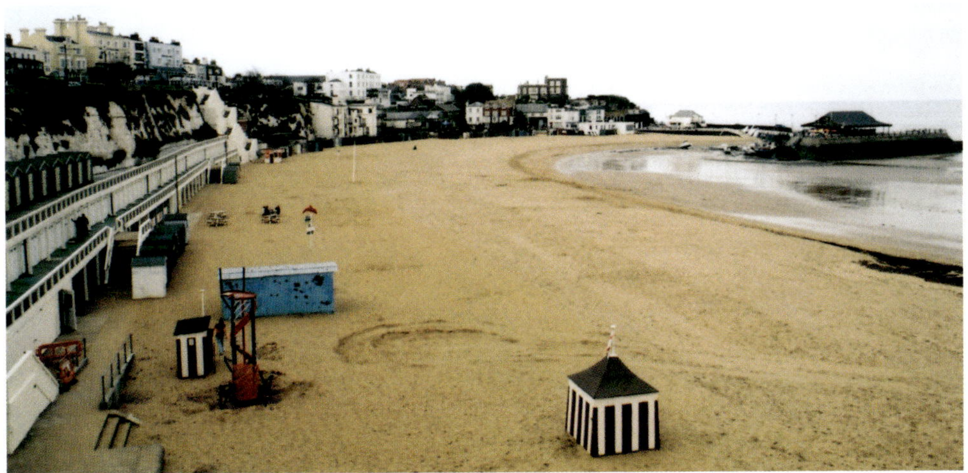

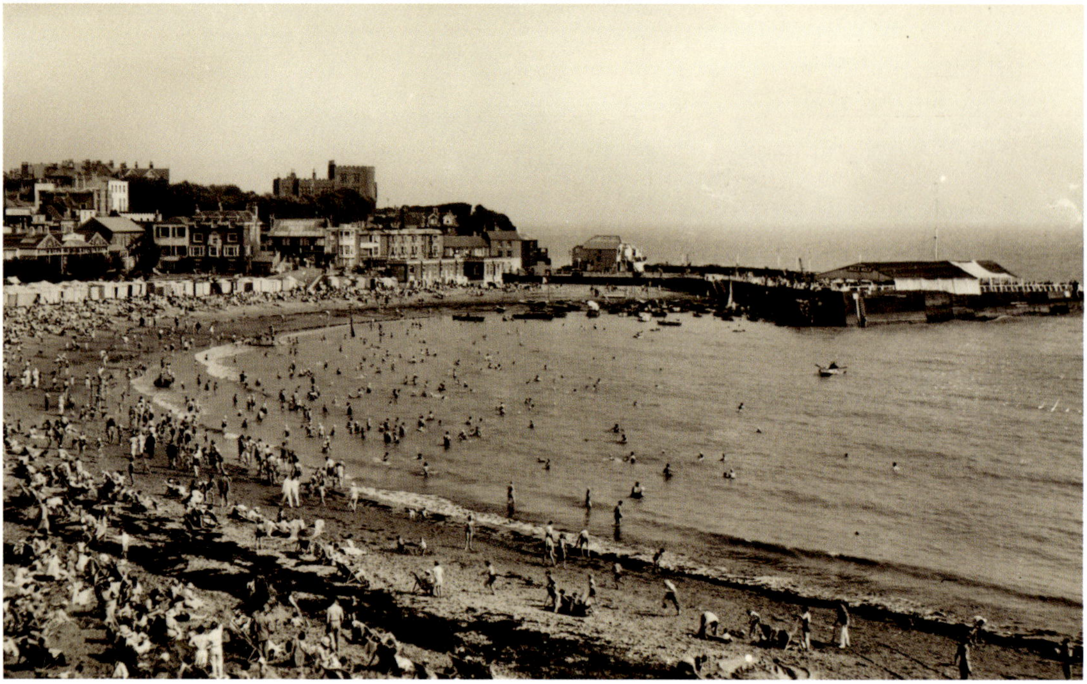

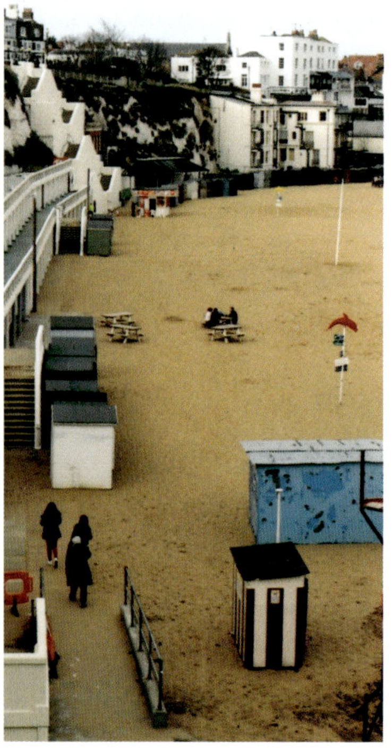

The Harbour

To the casual visitor the harbour seems to be an unchanging and attractive feature of the town, but for those who have grown up with it and who know it well, Broadstairs harbour is something that has always needed constant care and maintenance. It has been here for so long that nobody quite knows when it was first built. One of the earliest writers to mention it was John Lewis in his history of the Isle of Thanet, first published in 1723, but even he fails to offer a date. Broadstairs sits on top of the chalk cliffs overlooking the harbour, with a long, steep road, Harbour Street, passing through an old stone arch which leads to Viking Bay, the main beach. From the sixteenth to the mid-nineteenth centuries Broadstairs was well known for shipbuilding and the stone arch, York Gate, was built in 1538 as protection for that industry. Originally it was called Flint Gate and had two heavy wooden doors that could be closed in times of threat from the sea or privateers coming ashore, but in 1795 it was extensively rebuilt and renamed York Gate, after the 'Grand Old' Duke of York. Evidence of Broadstairs' maritime past can be seen as you pass under York Gate and come to the harbour, where on the jetty is the Old Customs House and seventeenth-century weather-boarded Boathouse. There are also many quaint and charming old fishermen's cottages and sheds, now tastefully converted into shops and restaurants. One such property is No. 9 Harbour Street, a flint cottage, now converted to a shop called 'The Old Curiosity Shop' with a wishing-well outside. Around 350 years ago this well was used by smugglers to hide their contraband.

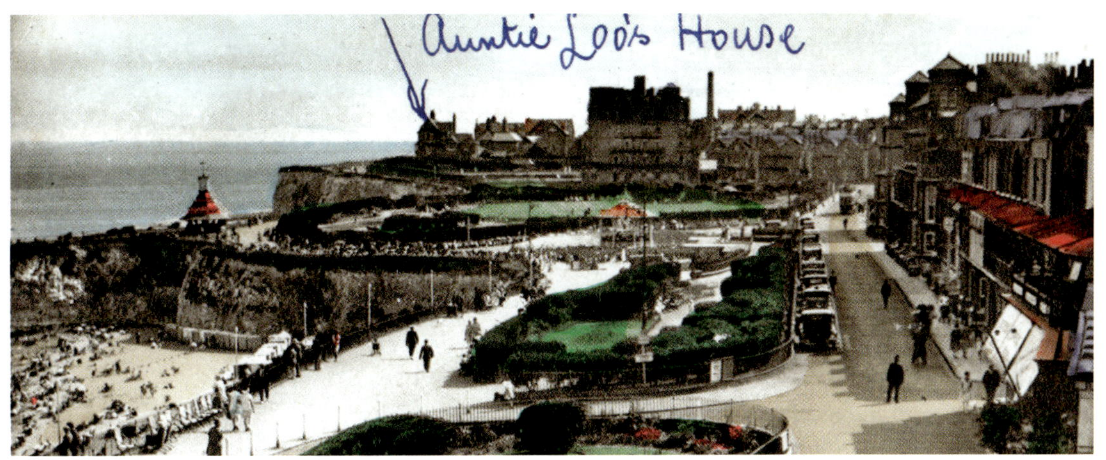

Broadstairs' Promenade

By 1910, Broadstairs' population had reached about 10,000 and a guidebook written by A. H. Simison, a local photographic chemist, in the 1930s described Broadstairs as 'having approached modernisation and urban development with a consistent policy of retaining those characteristics for which it has for so long been renowned'. Today Broadstairs is a magnet for visitors and has been likened to an archetypal Cornish fishing town; there are indeed many similarities. The clifftop promenade has a fine selection of hotels, houses and shops that are representative of each era from the late eighteenth century onwards, and from this early view it can be seen how all the ingredients of a successful seaside resort were in place – a promenade with sea views, a bandstand, formal gardens, a popular beach – but the question remains, who was the 'Auntie Loo' whose house is picked out? The postcard was stamped but not written upon, so we shall never know. Surprisingly for such a small town, Broadstairs has a clock tower, positioned on the cliff top overlooking the wide, sandy beach below. It is a more modest structure than those in other resorts, and has a circle of seating around it. Known as the Jubilee Clock Tower, it was built in 1897 for Queen Victoria's Jubilee celebrations. In 1949 a copper Viking ship weathervane was added to commemorate the landing of the Viking ship *Hugin* at Broadstairs in that same year, but in 1975 the tower was burnt down due to an electrical fault. It was rebuilt by apprentices at Thanet Technical College to commemorate Queen Elizabeth II's Silver Jubilee. Sadly, today the same view of the promenade is cluttered with parked vehicles, street furniture and growing foliage, and how the early photographer managed to capture such an elevated view remains a mystery.

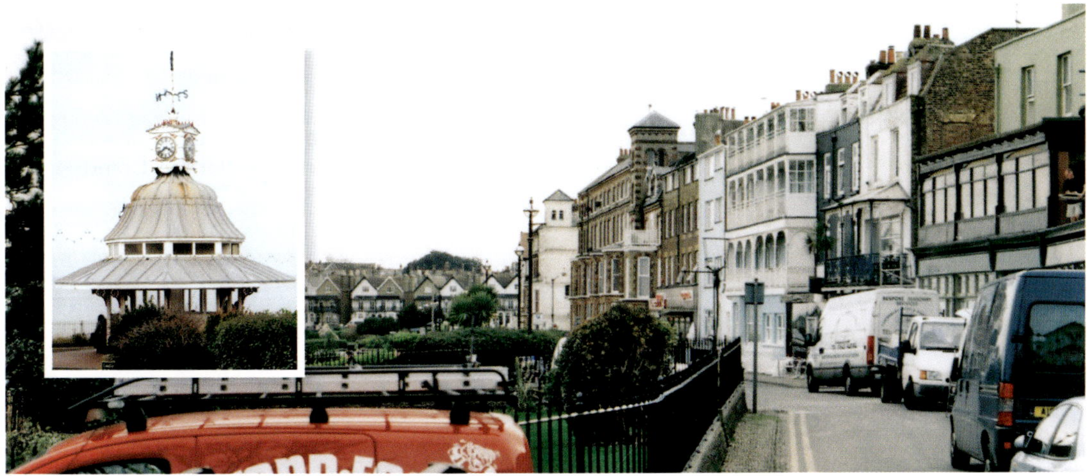

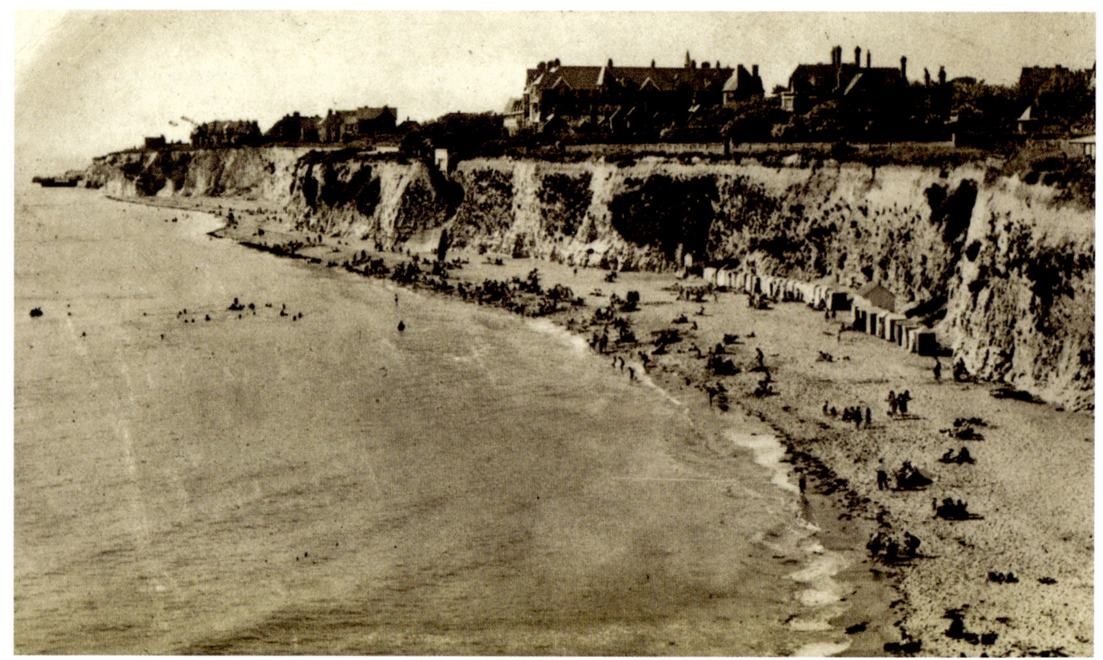

Stone Gap

Broadstairs consists of seven small bays where sunbathers and swimmers can indulge their passion. One such bay is Stone Gap, which lies to the east of the main beach, Viking Bay. It is accessible by steps down the cliff or along the promenade from neighbouring bays. As can be seen, it has no facilities whatsoever – just a fine, sandy beach.

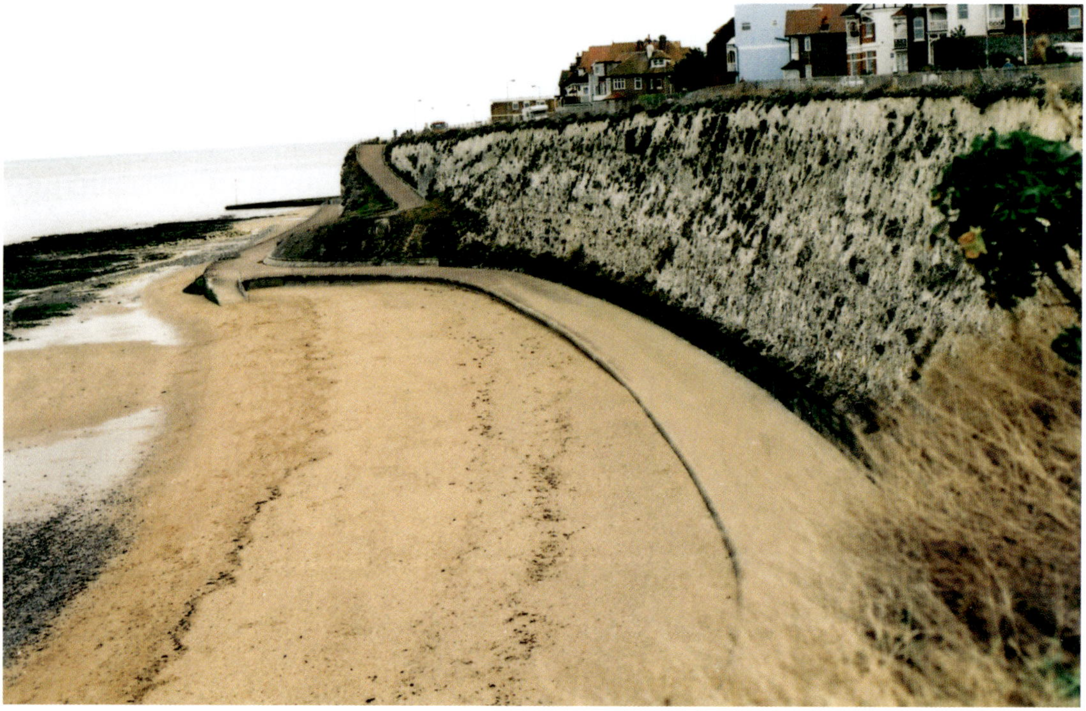

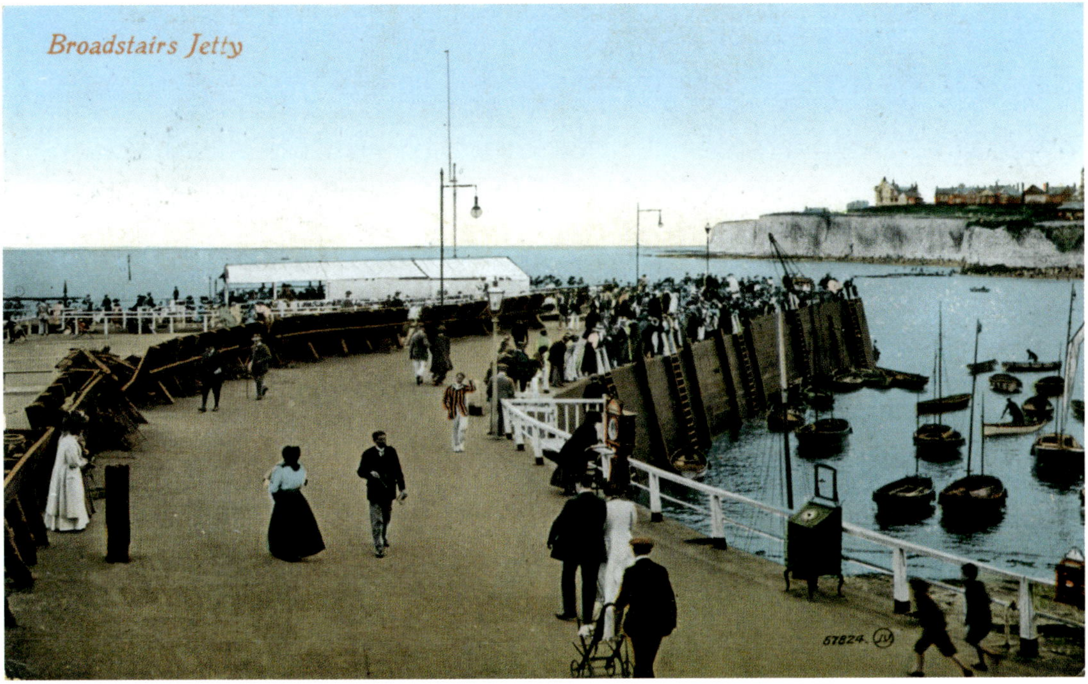

The Jetty

Being little more than a wall to shelter the harbour, the jetty has never had any delusions of grandeur. It has long been exactly what it is but is still nevertheless somewhere to stroll, be seen or watch the local fishermen, as can be seen in this 1911 view.

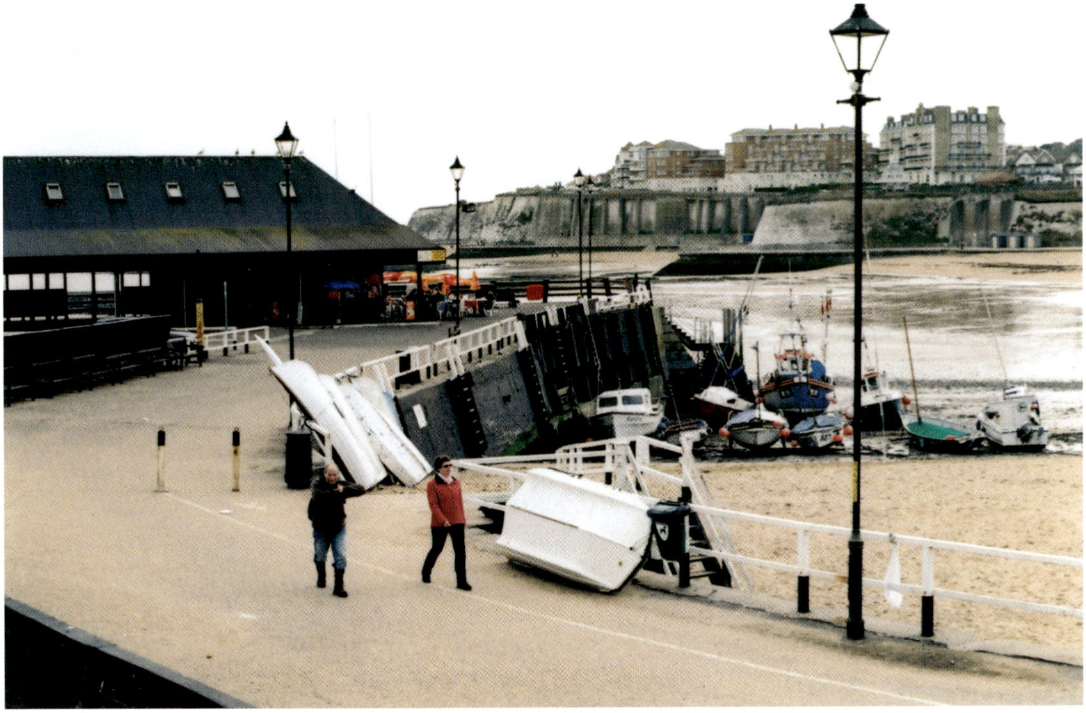

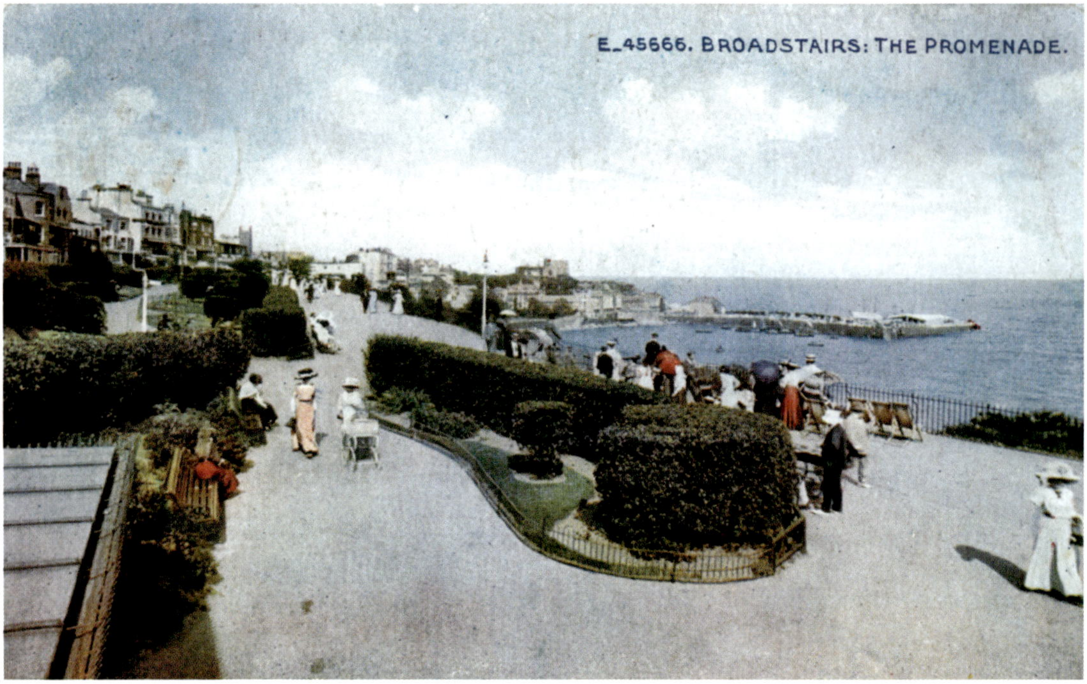

Promenading

Broadstairs promenade has changed little over the centuries, as can be seen in this 1922 view compared with that of today. The prom offers expansive views over the beach toward the harbour. I myself have spent much time here, staring out to sea and trying to see who was on the beach at the time.

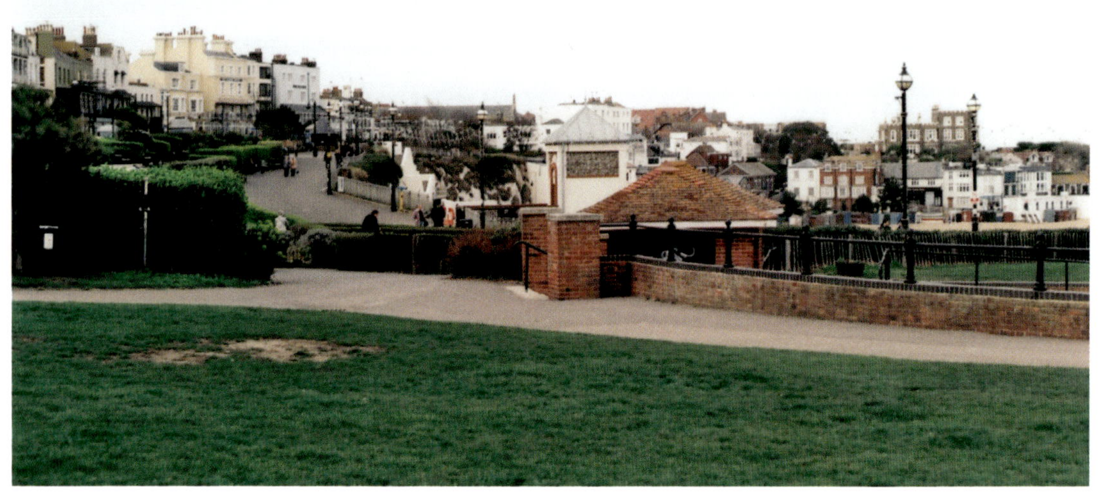

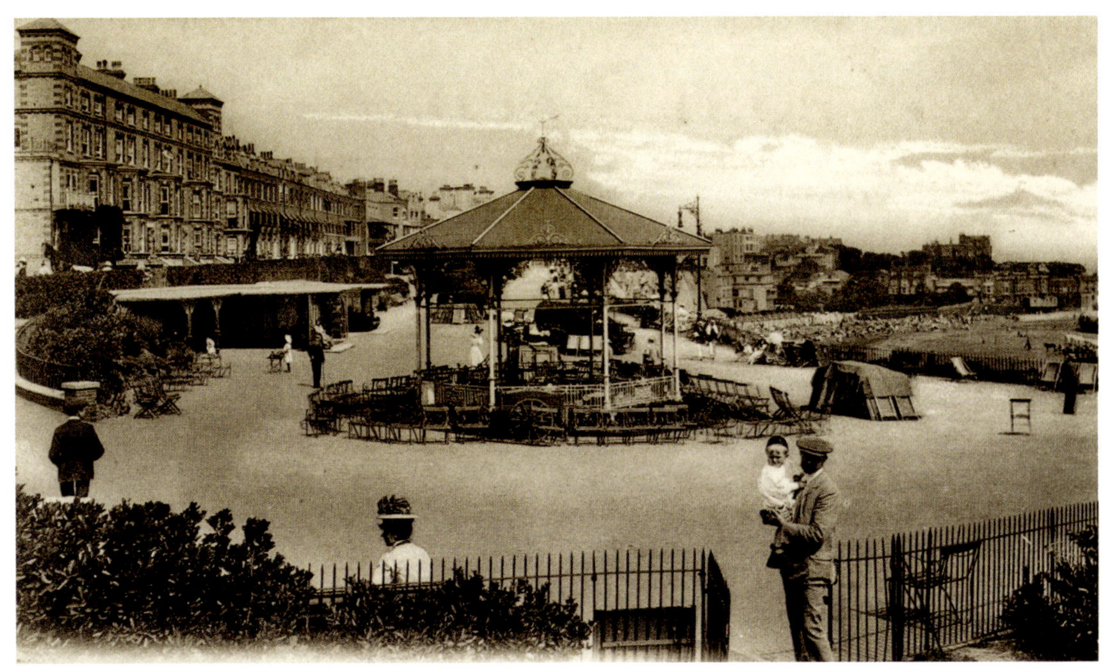

The Bandstand

In common with most seaside resorts across the UK, Broadstairs is also proud to have an historic bandstand. Broadstairs bandstand was erected during the Victorian and Edwardian eras, when musicians would play lively band music to tourists strolling along the promenade and looking out to sea across Viking Bay. This tradition continues throughout Sunday afternoons between 2 and 4 p.m. in the summer months. The bandstand has changed very little, as can be seen in this view of 1908 compared with that of today.

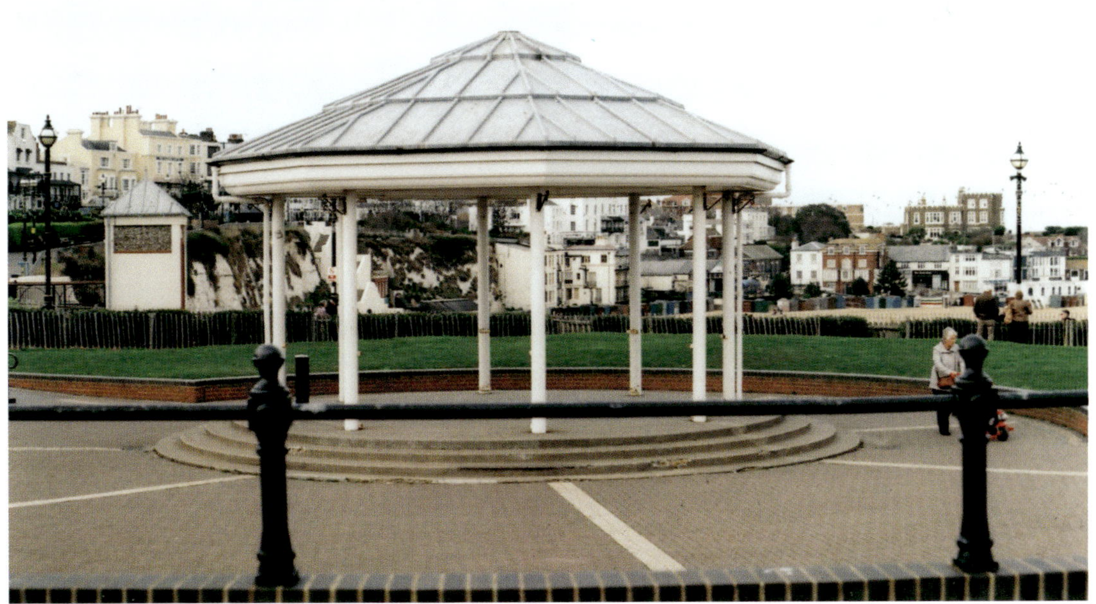

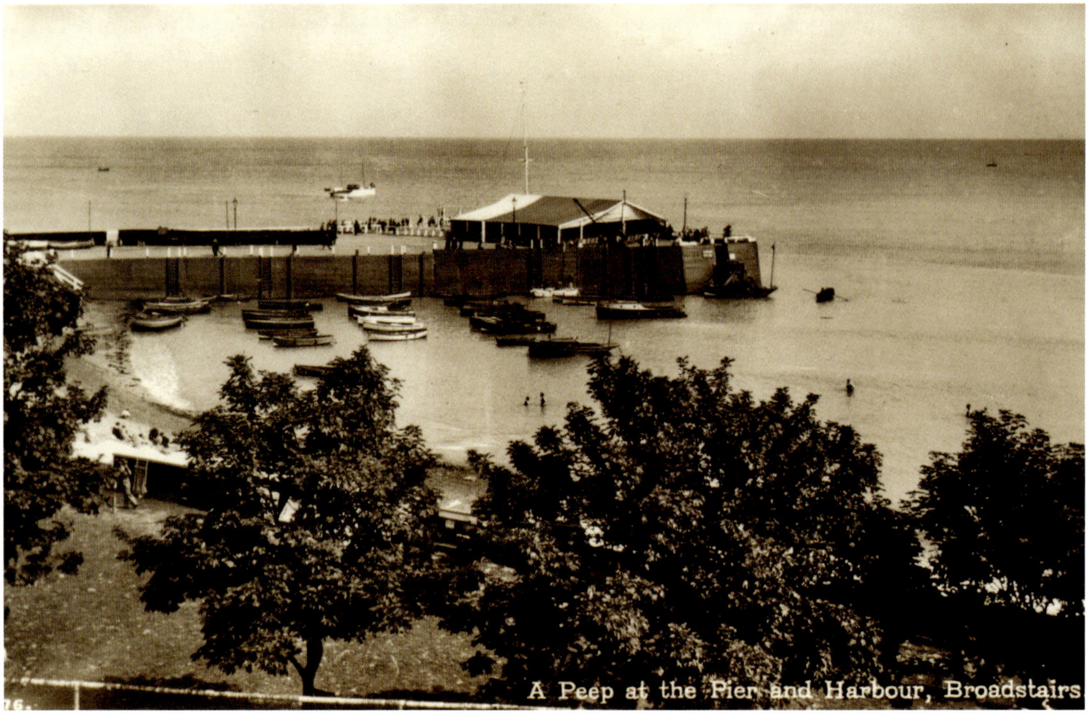

The Jetty from the Promenade

Broadstairs harbour is still very much in daily use, the only difference being that the commercial fishing vessels seen in earlier views have now been replaced by recreational craft.

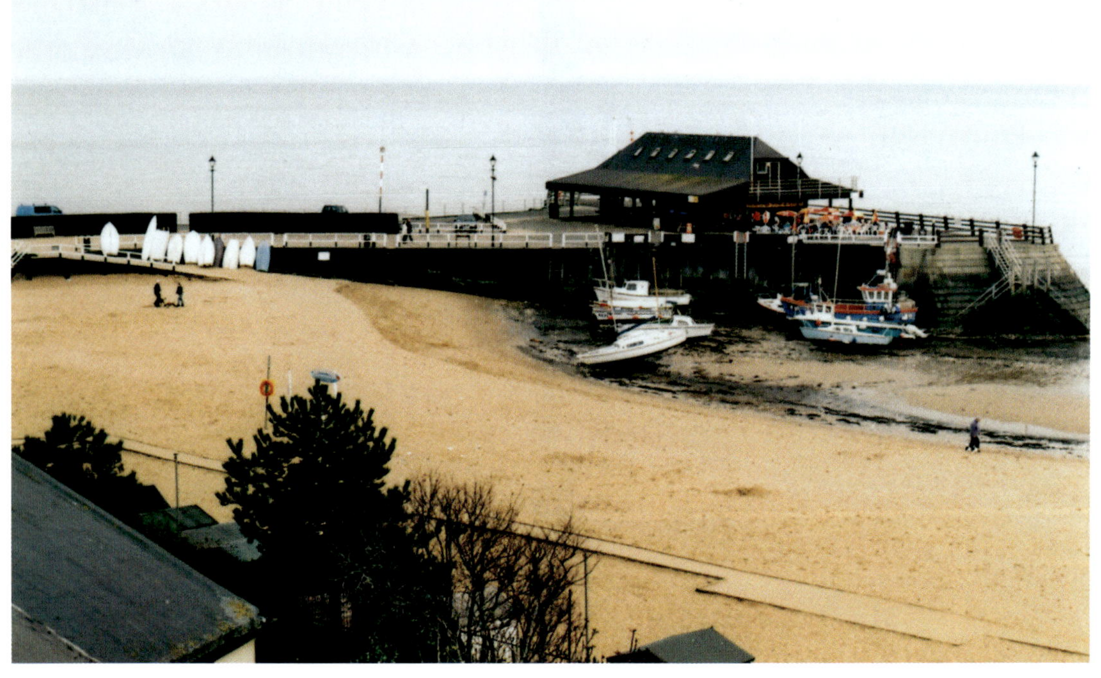

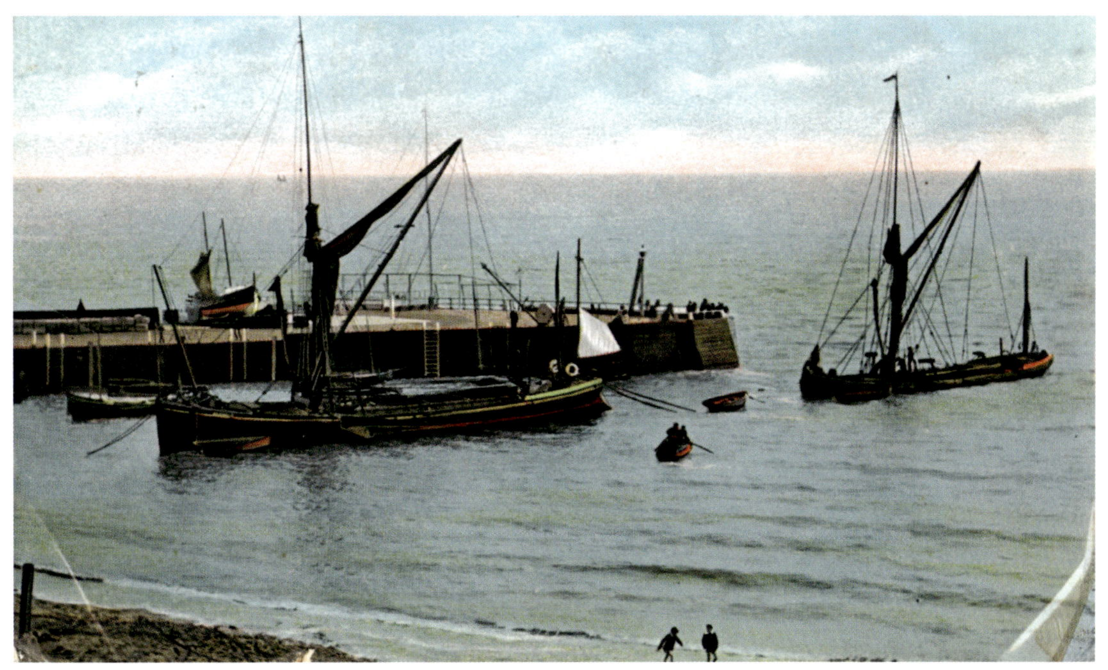

Entertainment on the Jetty

This 1907 view of sailing vessels moored in the harbour caught my eye, not so much for the picture but the message on the back, sent to a lady by a lad named Jack. It reads, 'This is what they call a pier. It only consists of a few planks of wood and a shelter. The band plays on here in the morning and n*****s perform in the evening. Jack.' Derisory or what? I do not know to which band he was referring but 'the other entertainment' was most likely Uncle Mack's Broadstairs Minstrels, who regularly entertained both locals and visitors from 1895 to 1948.

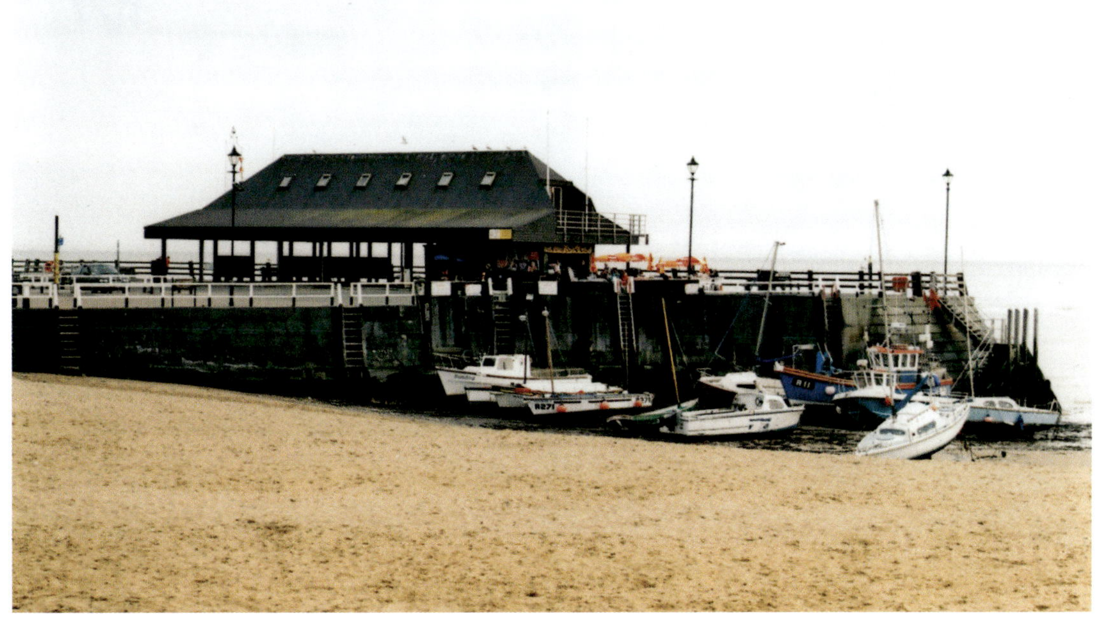

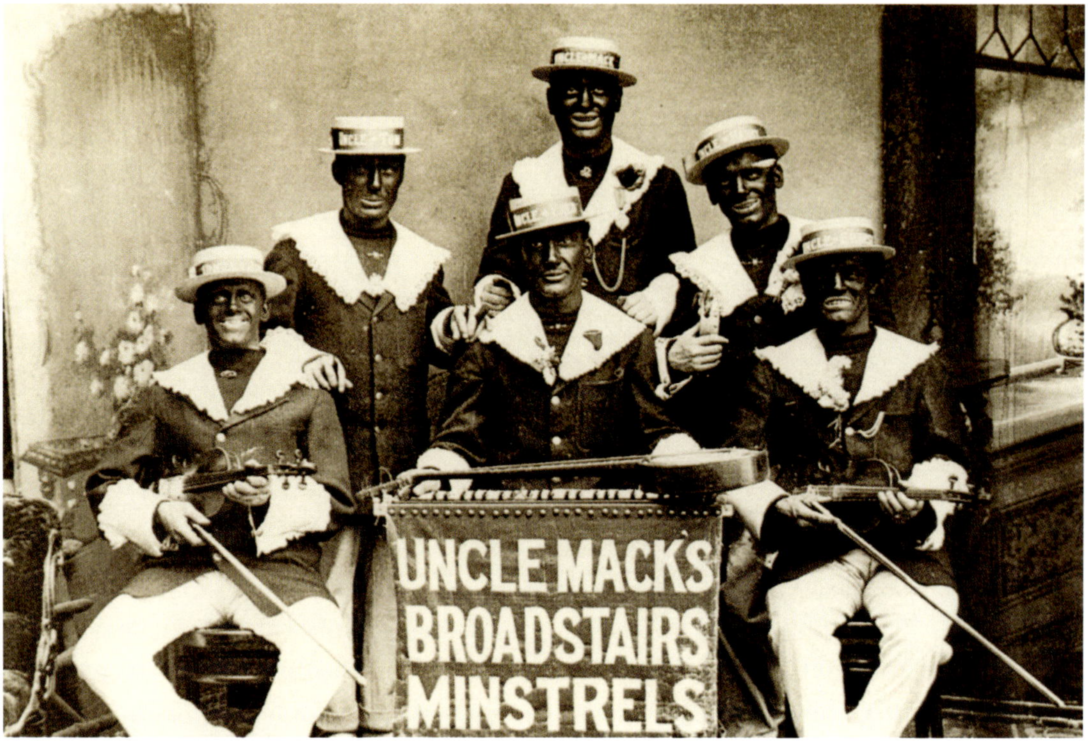

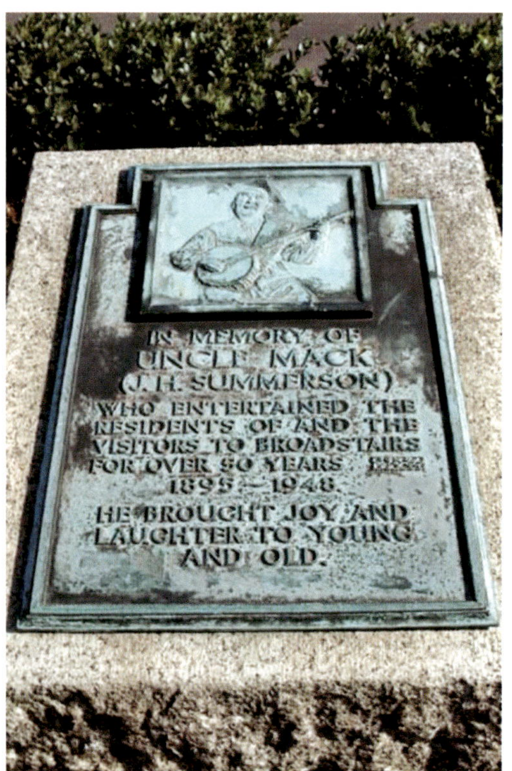

Uncle Mack's Broadstairs Minstrels

Uncle Mack's Minstrels were an integral part of the Broadstairs summer entertainment in the early part of the twentieth century from 1895 to 1948. In 1911, they were voted the most popular troupe of the British seaside resorts. Now considered unacceptable as a form of entertainment, black-faced entertainers were a popular act in the 1930s and were often seen at seaside resorts, where the additional character of 'Uncle' befriended children and encouraged them to take part in singing and dancing competitions. 'Uncle Mack's' real name was J. H. Summerson; the rest of the troupe were Alex, Jack, John, Harry and Barney. Their song-and-dance routines were accompanied by piano, drum and banjo. Changing attitudes to race and racism led to the Race Relations Act of 1976, which stopped 'black-face' entertainments in the UK in the early 1980s. Such was the high esteem in which the residents of Broadstairs held Uncle Mack that a memorial dedicated to him was placed on the promenade and unveiled at Whitsun 1950 by the actress Annette Mills, of Muffin the Mule fame.

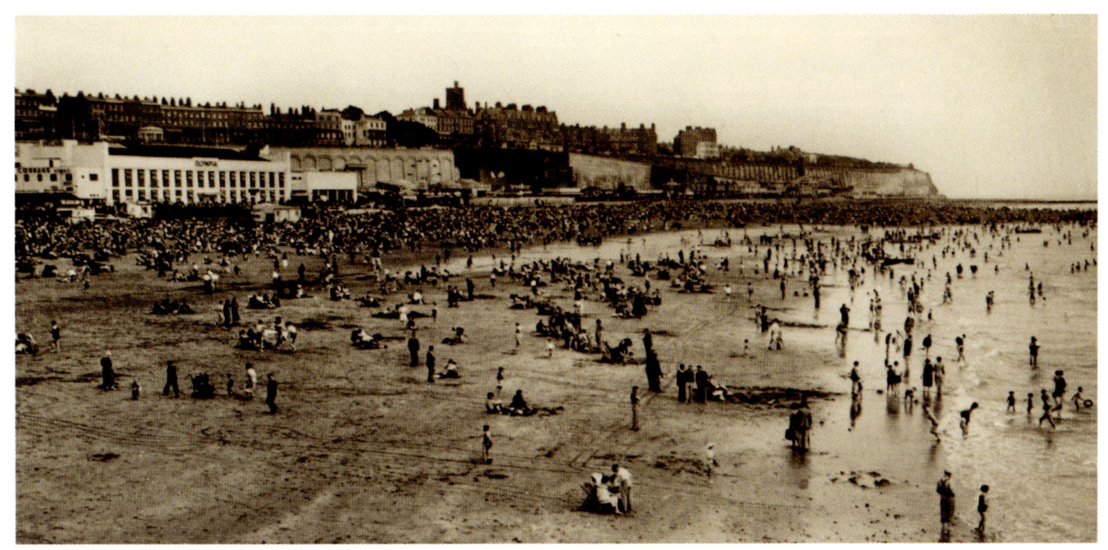

Ramsgate

A few miles on from Broadstairs, as the Thanet coastline drops southward into Pegwell Bay, is Ramsgate, with its extensive beach and large harbour. Initially, Ramsgate was just a collection of fishermen's cottages near the harbour but later on, towards the end of the seventeenth century as shipping trade increased, the port began to grow in importance. In the middle of Ramsgate, four major roadways crossed where the centre of the town now is and along which houses were gradually built. These were elegant Georgian houses, with the beautifully proportioned sash windows built throughout the eighteenth century, later evolving into the bow-window-fronted houses of the Regency period, 1812–20. The eighteenth-century craze for sea-bathing contributed greatly to the development of Ramsgate, and Queen Victoria, when still a princess, stayed in the town on several occasions in the 1820s, helping to make it even more fashionable. She later bought the famous painting *Ramsgate Sands* by W. P. Frith, perhaps as a memento of happy childhood days. The coming of the first railway in 1846, with a station on the sands, brought ever-increasing numbers of visitors to the town.

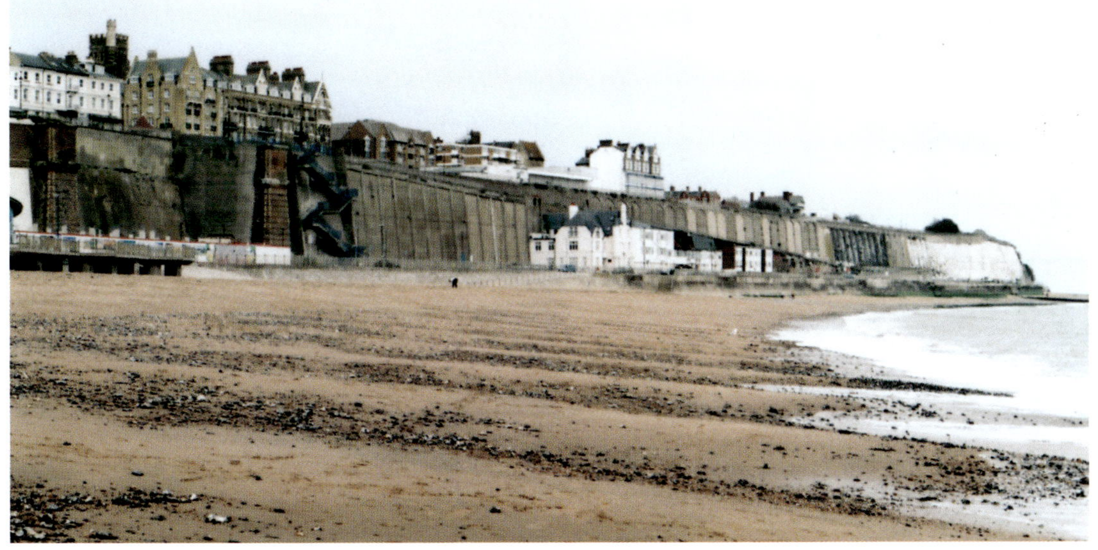

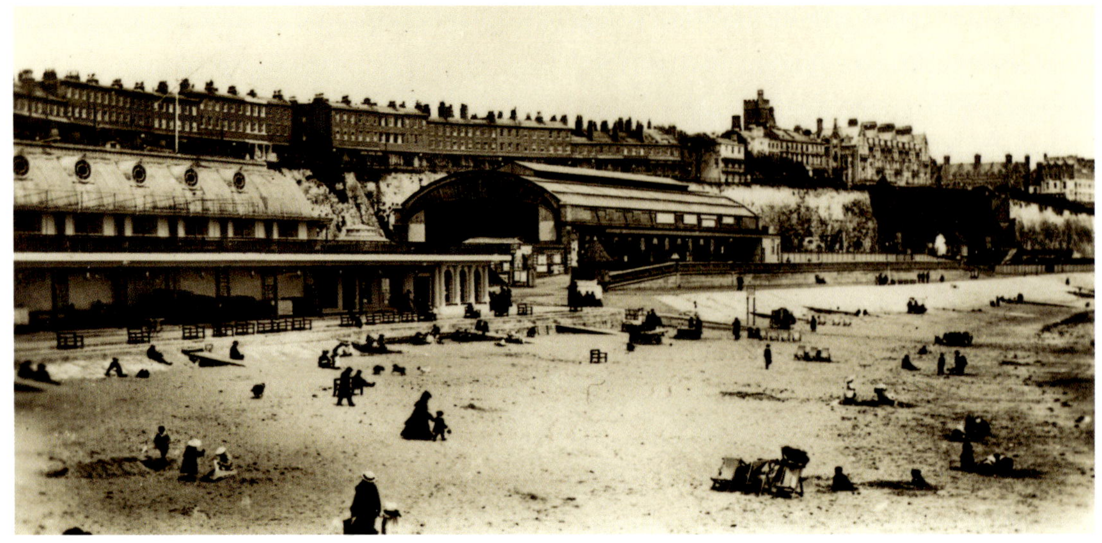

Ramsgate Sands Station

Those who have journeyed to Ramsgate by train will know that the present station is situated just outside the town and is a long walk from the harbour and beach. It has not always been so. The railway came to this part of Kent in stages. First it terminated at Margate, with a terminus on the site of Dreamland. Later it was extended to Ramsgate, terminating by the beach, and finally the line was extended to Sandwich. On each occasion it meant a new station had to be built on the line of the extended railway, which is why in some cases no consideration seems to have been given to convenient access to the seafront. The railway first came to Ramsgate in 1863 and at that time the line came to Dumpton Park before entering a long tunnel in the cliff, emerging on the seafront and into a terminus station. The terminus closed in 1926 when the line was extended from Dumpton Park to Sandwich and a new station, the present one, was built. The old terminus was then sold to become a funfair, Pleasureland, and later, during my childhood, Merrie England. The wooden building caught fire and burnt down in 1998, and despite plans to redevelop the site being announced, nothing ever came of it. The site was cleared and today is a car park. The inset picture shows a steam train leaving the terminus, heading for the mouth of the tunnel in the cliff, which even to this day is still discernible.

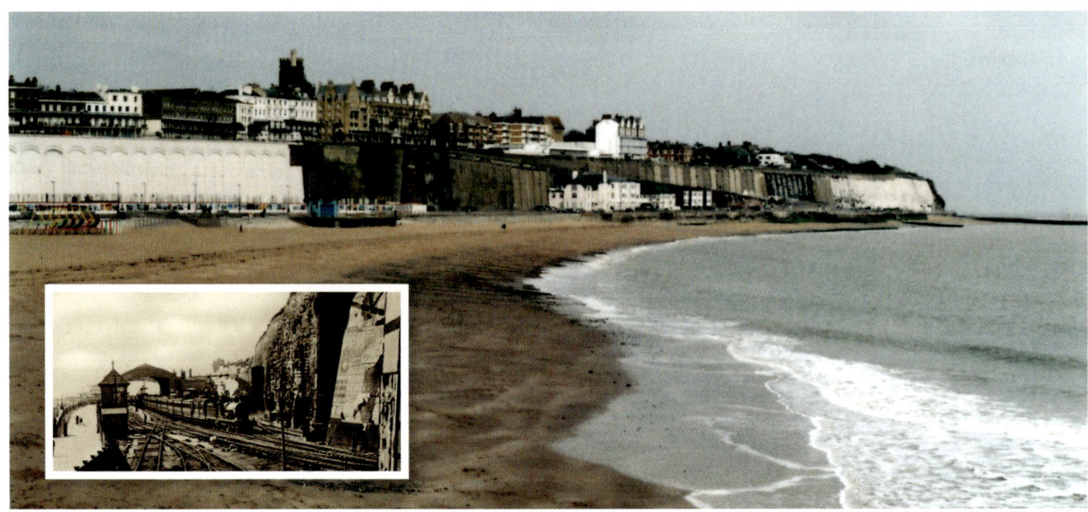

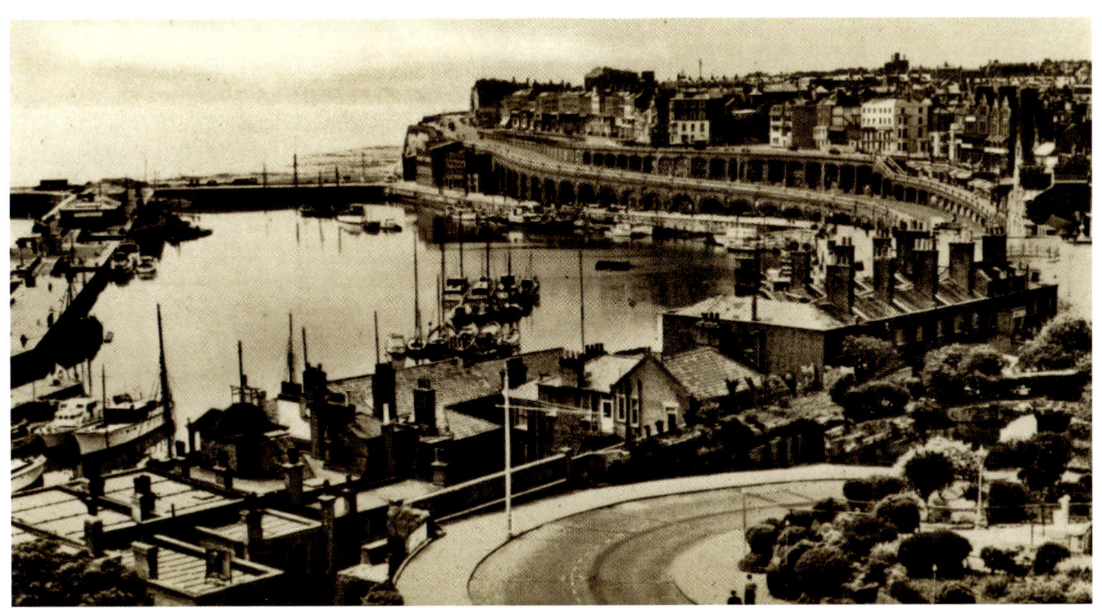

The Royal Harbour

The harbour was built as a result of a Great Storm in 1703 which saw the loss of much shipping. Two hundred ships anchored in the Downs, a channel between the coast and the Goodwin Sands, were blown about by winds reaching 120 mph. The majority of the vessels were wrecked and 1,200 people lost their lives. Construction work on the harbour began in 1749 and was completed in around 1850. The harbour has the unique distinction of being the only harbour in the UK to be awarded the right to call itself a Royal Harbour, bestowed upon it by King George IV who was much impressed by the hospitality shown by the people of Ramsgate when he used the harbour to depart and return with the Royal Yacht Squadron in 1821. A granite obelisk in Pier Yard commemorates this event. Because of its proximity to mainland Europe, Ramsgate was a chief embarkation point during both the Napoleonic Wars and the Dunkirk evacuation in 1940. Today the harbour is given over to small fishing vessels and yachting marinas in both the inner and outer harbours while a new port, constructed outside and to the west of the old Royal Harbour, handles ferries operating daily sailings to Ostend in Belgium.

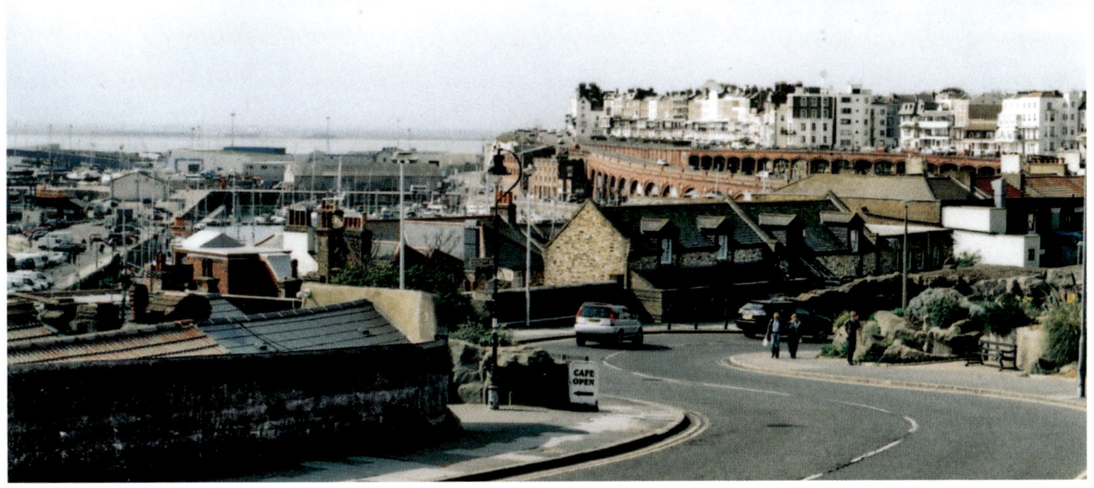

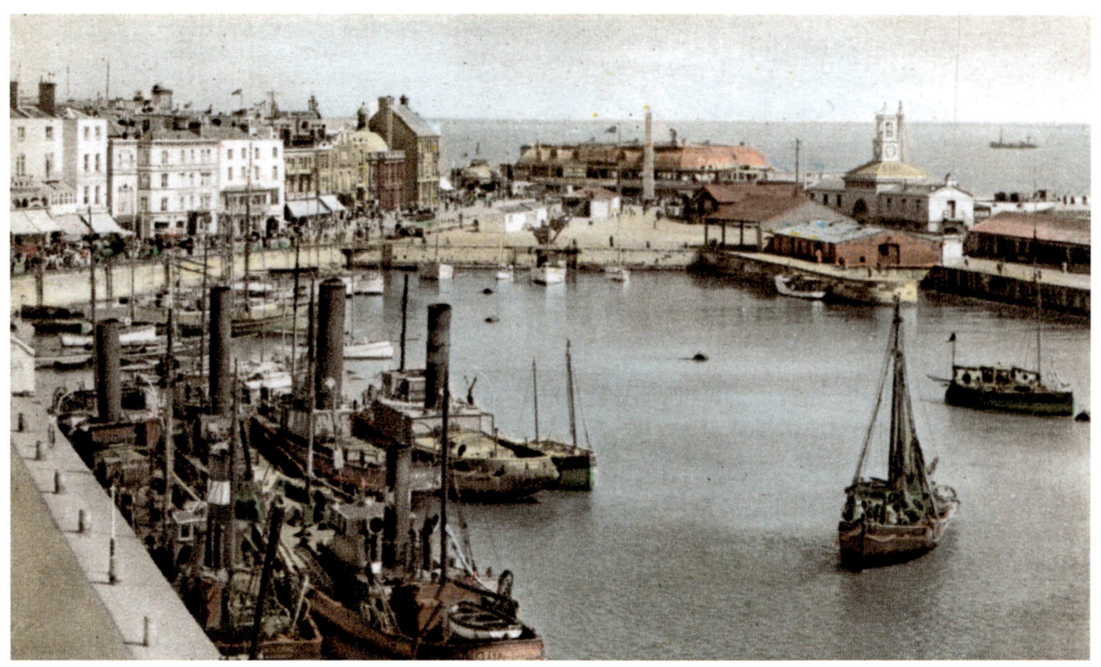

The Harbour Quayside

This was where the town of Ramsgate began, as a community of fisher folk living on the seashore. This they had done for many centuries and it was not until 1703 that a proper harbour was built. In this early 1900s hand-tinted view of the harbour it can be seen how working vessels dominated the harbour, whereas today it is mostly pleasure craft.

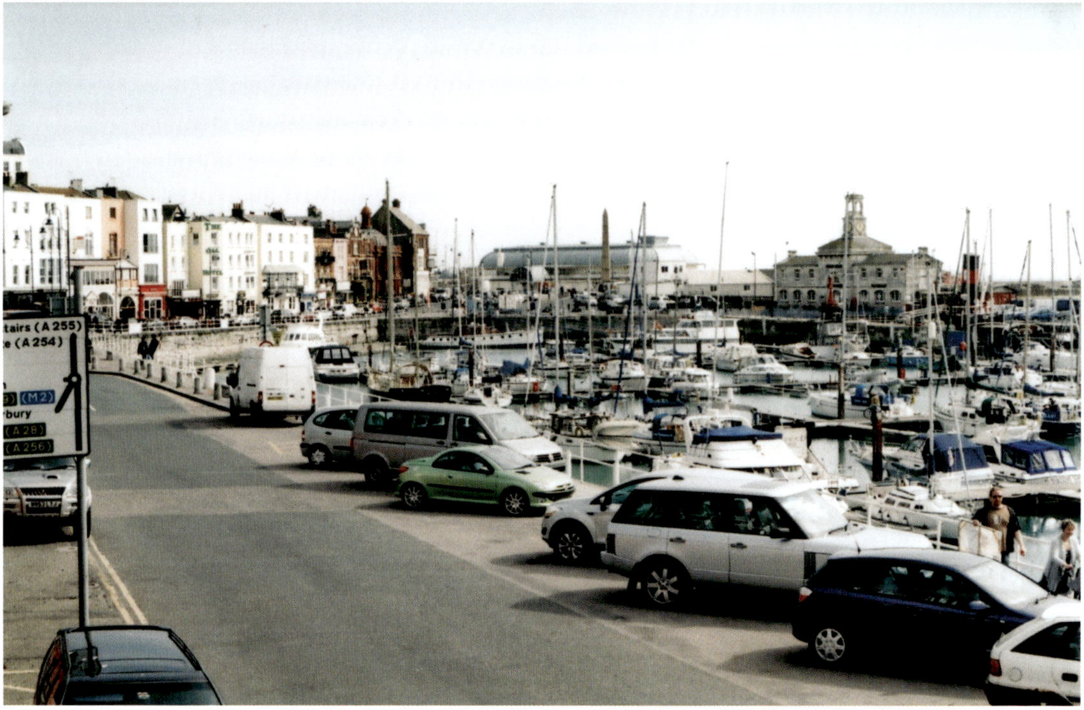

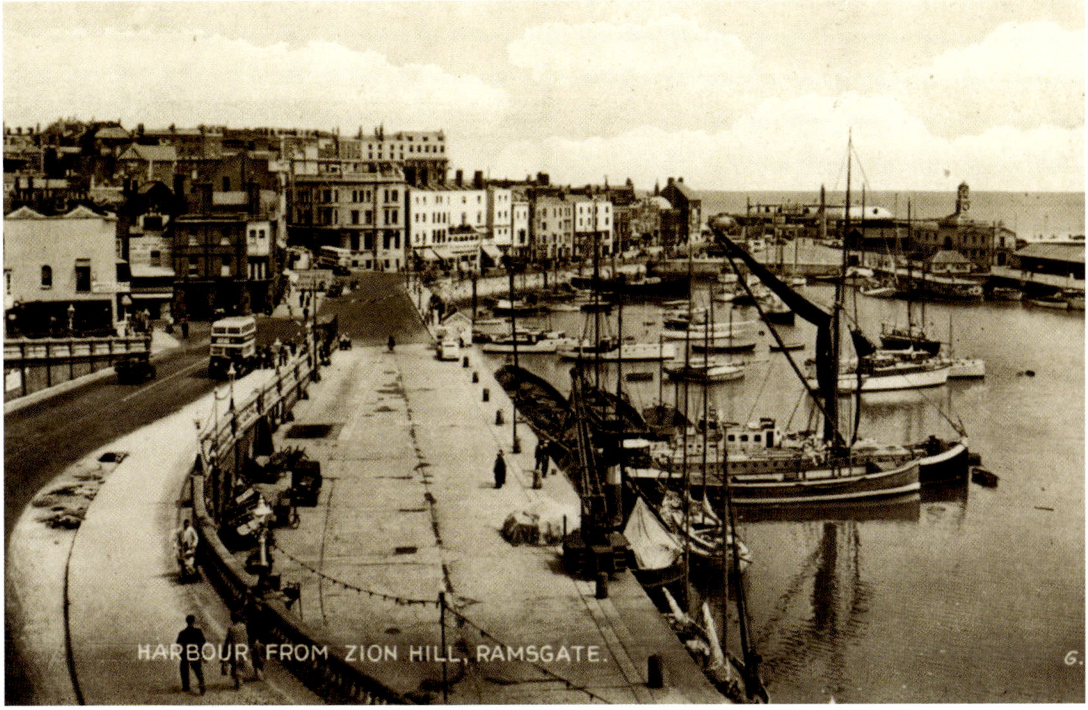

The Harbour from Zion Hill

The approach into Ramsgate is via Zion Hill, from the top of which visitors get an expansive view of the harbour and quayside.

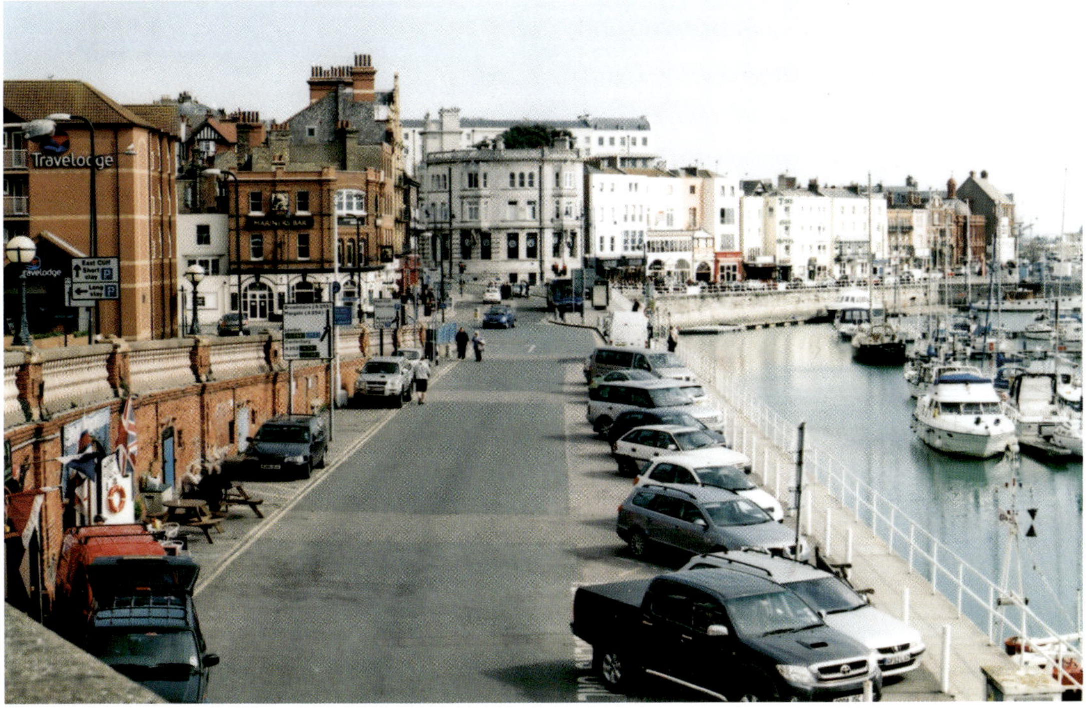

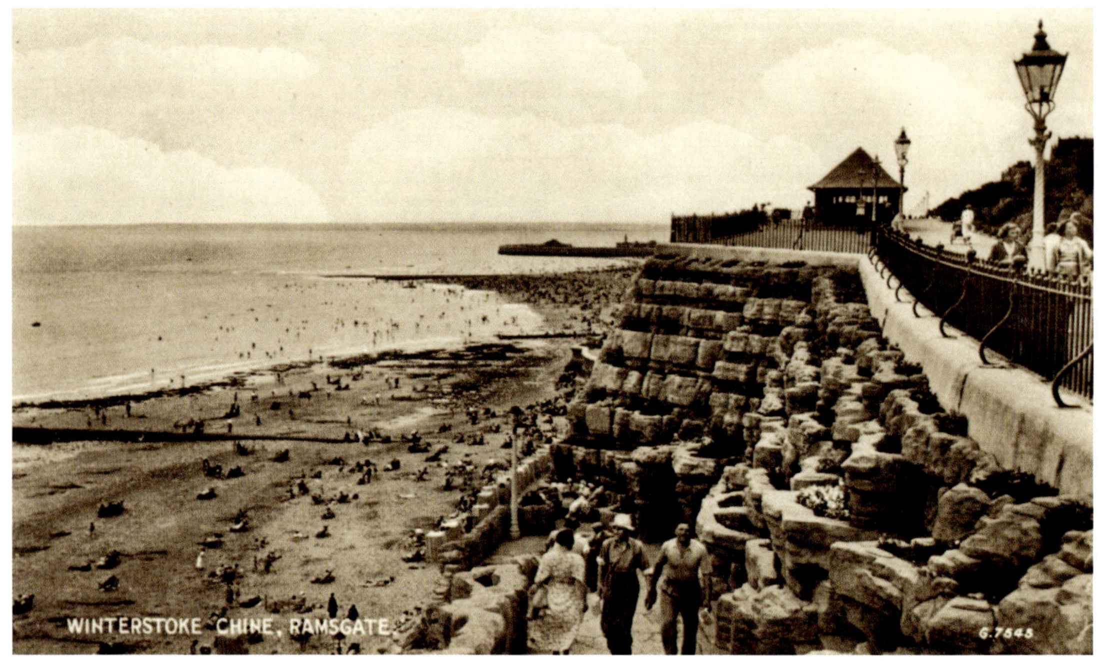

Winterstoke Chine

This was once a popular walk down a cliff path from the hotels and guest-houses in Victoria Parade, high on the cliff top, to the northern end of the beach below. It also gave direct access to the Marina Bathing Pool. Today's view is still very similar to the earlier one; little seems to have changed, except that the bathing pool has now gone.

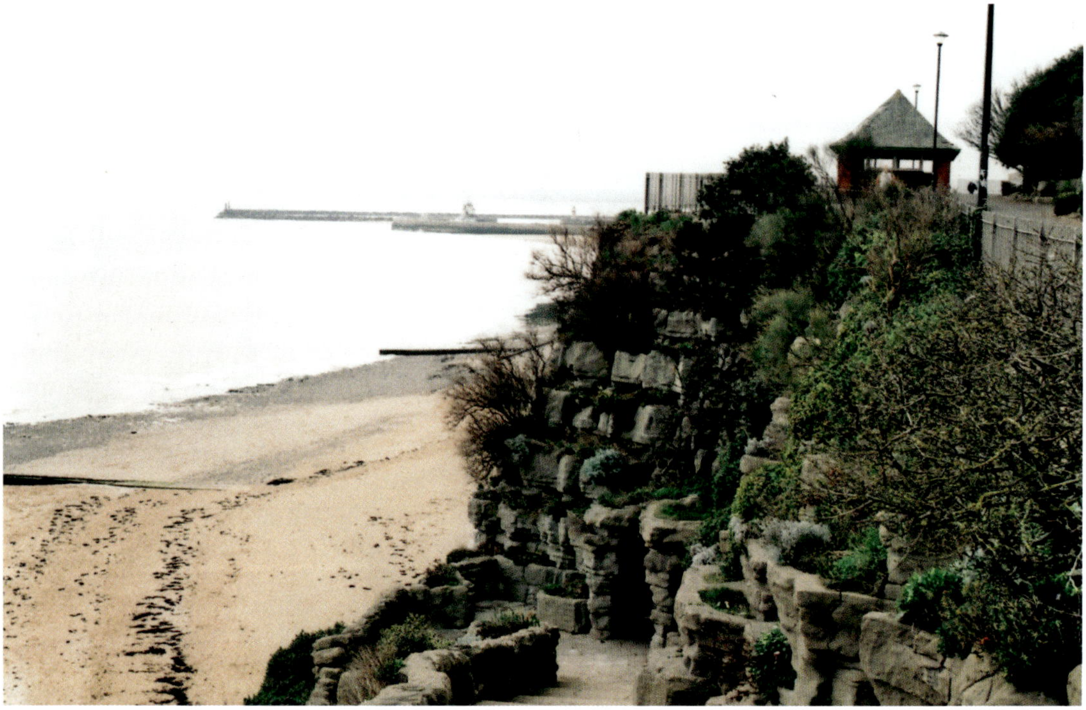

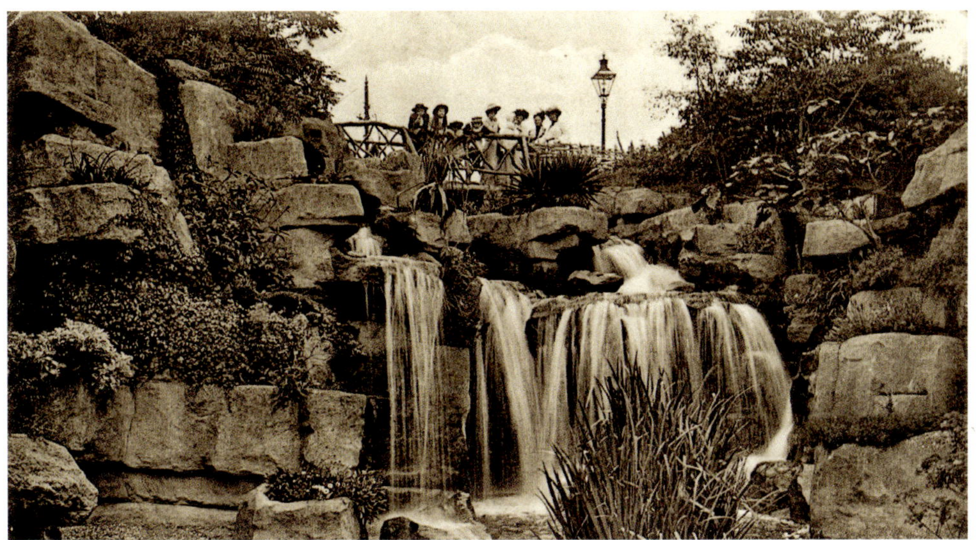

Madeira Walk

Another popular walk with holidaymakers was through Madeira Walk, which took them from the cliff-top guest-houses and hotels in Wellington Crescent, down to the harbour and its quayside area. This 1911 view shows there was much to be seen as you made your way down the Walk, with alpine-like flower settings and a splendid waterfall, much admired by this group of Edwardian ladies. Today, even though the setting remains, there is no longer any trace of the waterfall. As a garden, Madeira Walk was left to fall into ruins over the years but in 2011 the council, helped by the Ramsgate Society, began work on restoring it to its former beauty. This area is listed because it is home to one of the best examples of Pulham artificial rock in the country, but over the years self-seeded trees have obscured this and started to damage the rock. The council has now removed these trees, stumps and roots. The work also uncovered some of the original lights, which have now been put back into working order.

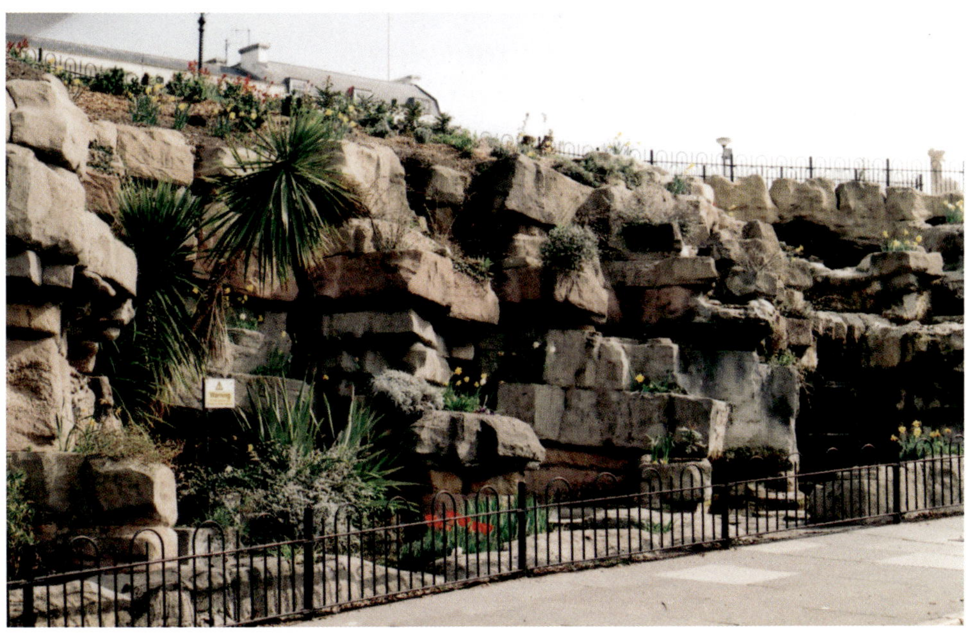

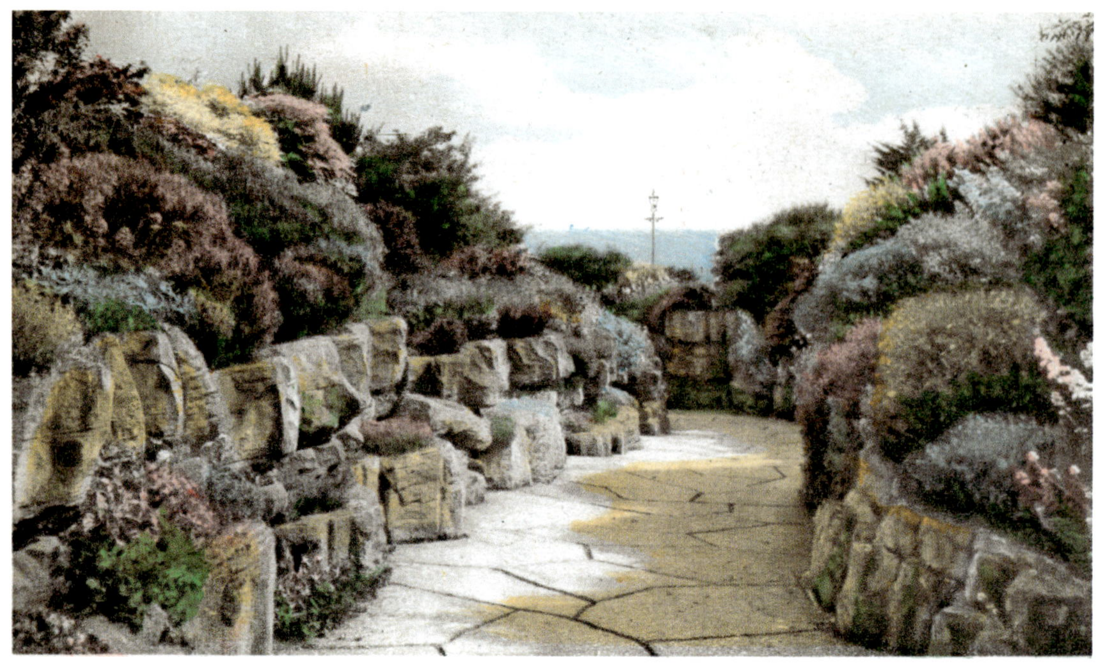

West Cliff Chine

Another pleasant walk from the cliff-top hotels and guest-houses in Royal Esplanade is through West Cliff Chine to the west of the harbour where, again, imaginative use has been made of Pulham artificial rock and alpine planting.

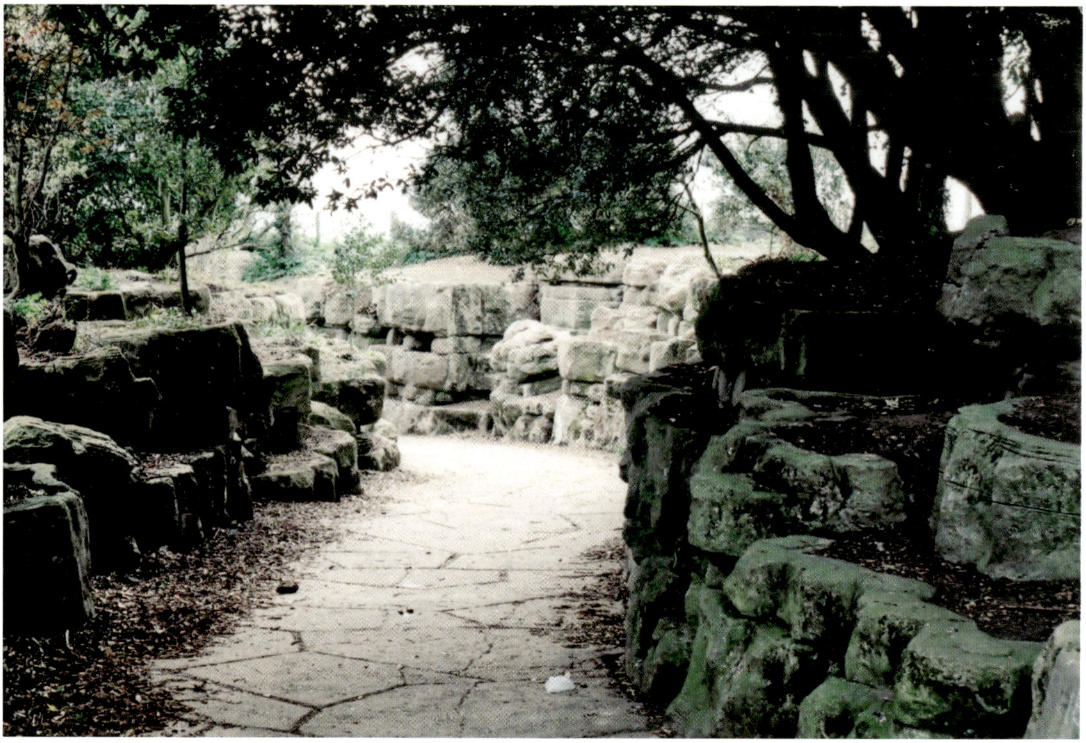

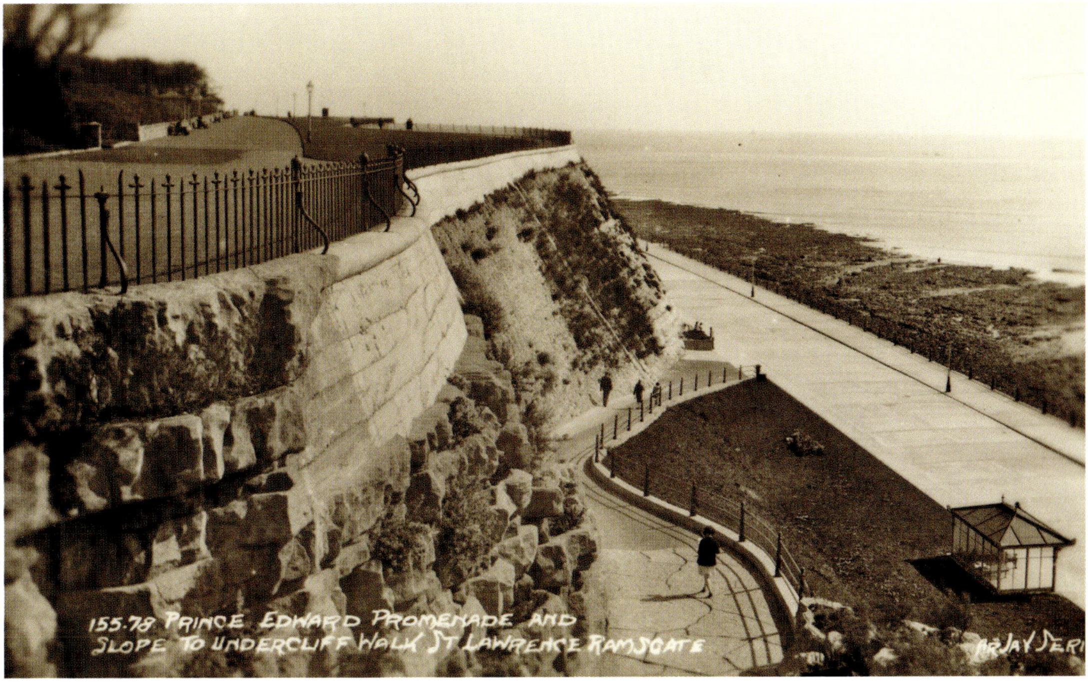

Prince Edward Promenade

Although less ornate than Madeira Walk and West Cliff Chine, this was nevertheless a popular way of getting from the cliff top to the rocky foreshore below to the west of the harbour. This undercliff walk would take visitors on a short stroll to Pegwell Bay.

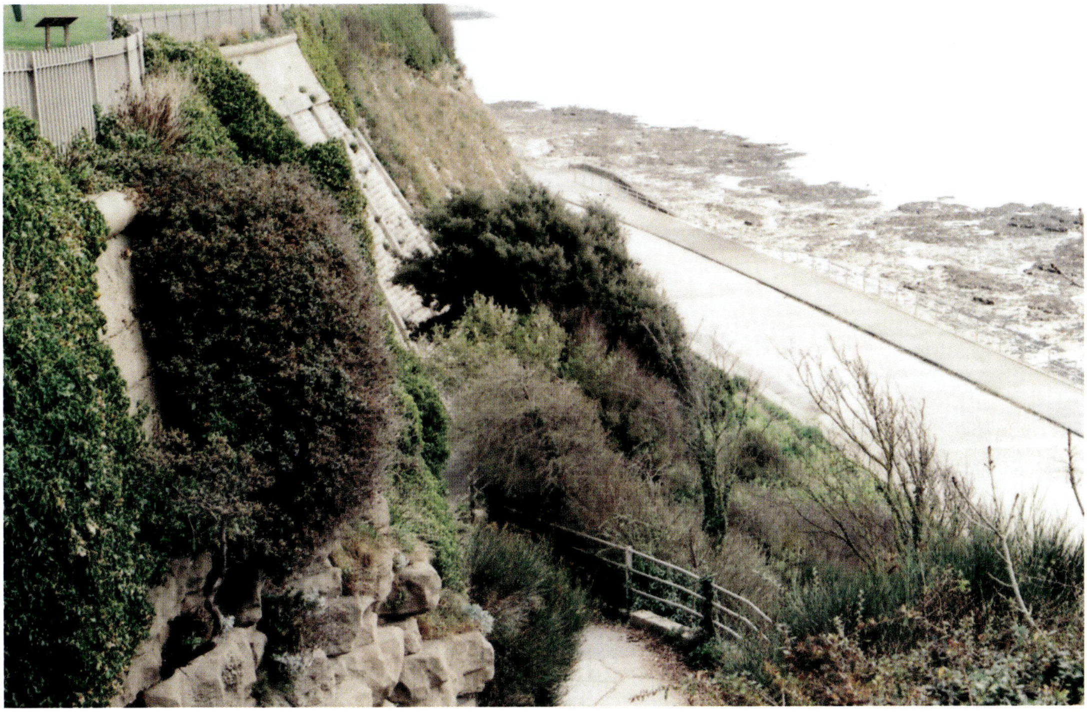

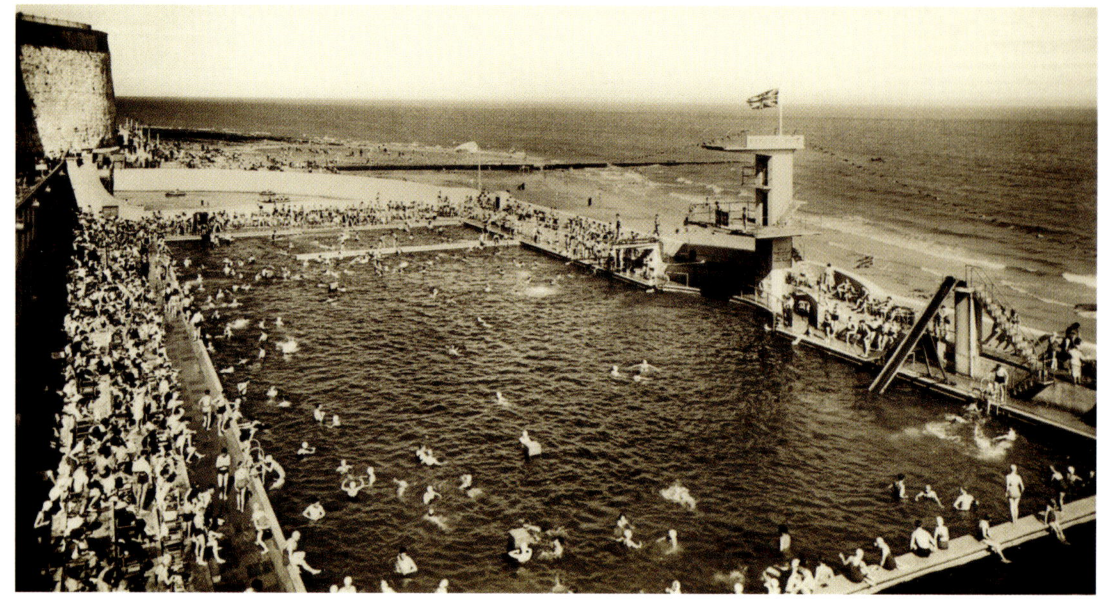

The Bathing Pool

Ramsgate, like most other seaside resorts, once had a grand outdoor bathing pool located on the beach, as can be seen in this undated view. The Ramsgate Marina Bathing Pool, to give it its full name, was opened in 1935 and was one of the largest outdoor bathing complexes in the UK; it was a huge attraction for visitors. Today it is nothing more than a large car park that the council is trying to sell off for its development programme. Since its demolition, however, it has become something of a magnet for the locals who want to find the underground structures that were once a part of the Marina Pool. These included changing rooms, shower rooms, locker rooms and more but it is more than likely that this underground part of the complex was filled in when the car park was built. Some people claim to have found a way in and have photographed what they believe to be evidence – or is it the remains of Nero's nightclub, which later occupied the site?

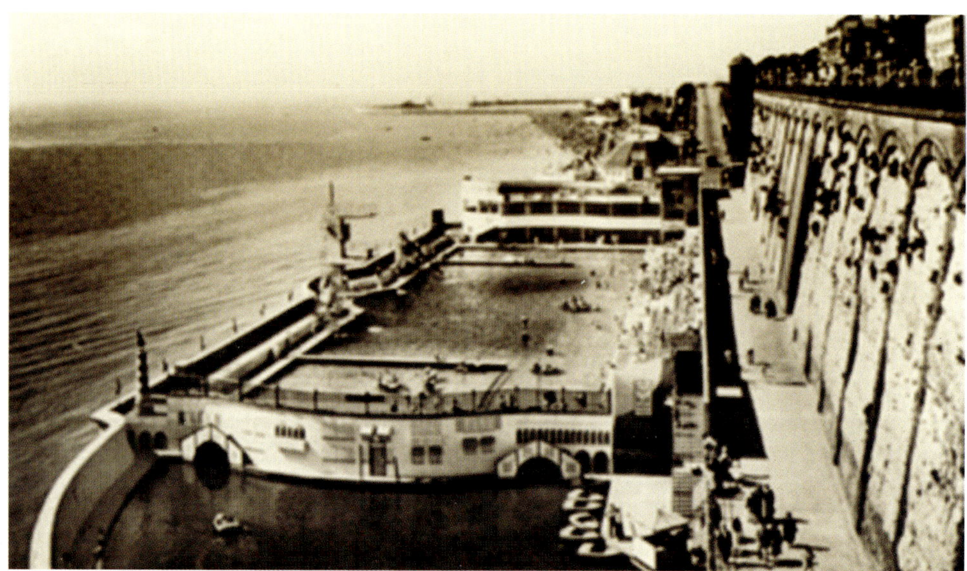

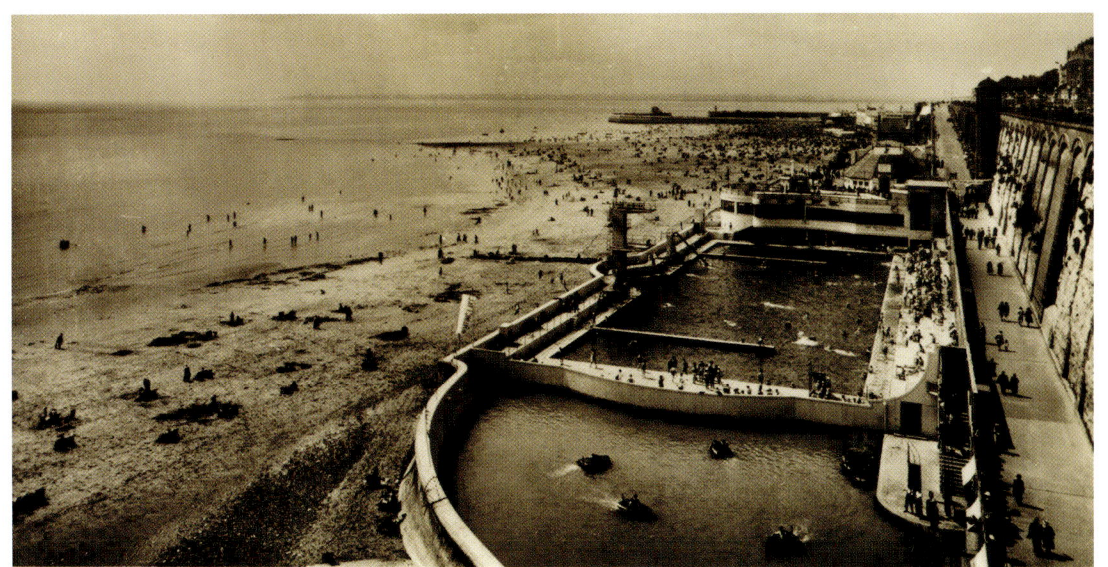

The Cause of its Demise

What caused the demise of the Marina Bathing Pool? Some say it was a council blunder that damaged the swimming pool so badly it could never be repaired and used again. The suggestion is that when it was decided to hand over the running of the pool to the leisure department, a set of instructions about running and maintaining it were also given by the engineers' department. One of the most crucial instructions was that the pool was never to be emptied without supporting the seaward-facing side first. The pool had been designed so that the weight of the water inside counteracted the forces of the sea battering against it. The first thing the leisure department did when they took it over was to empty it unsupported. In the resulting cover-up, councillors were told that the damage was caused by bombs falling during the Second World War that had weakened the structure.

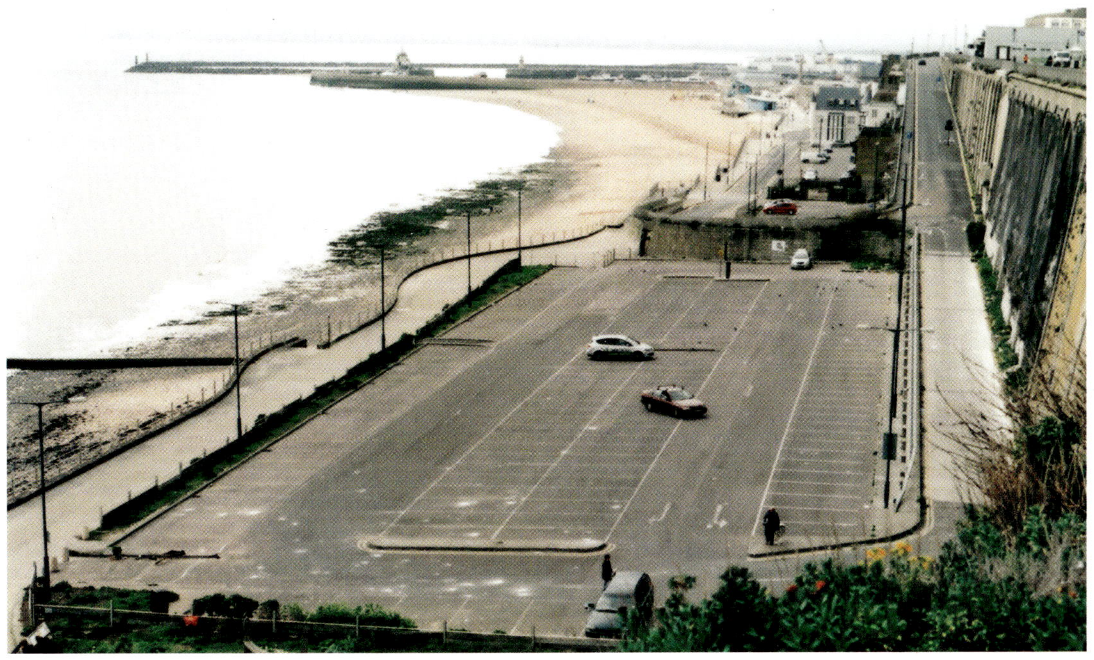

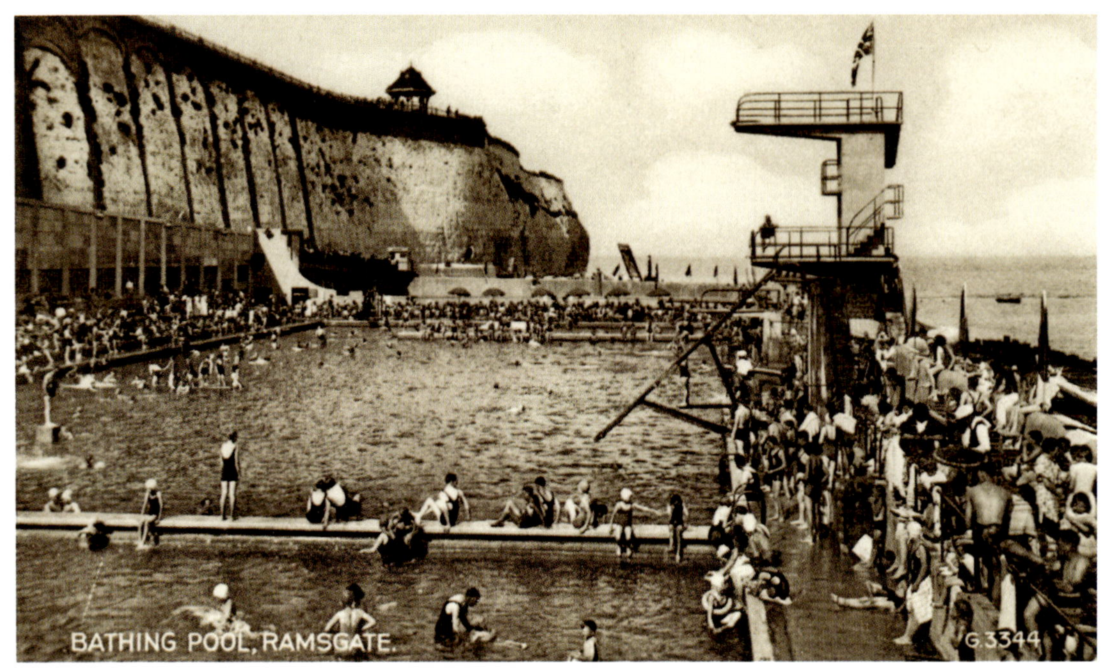

BATHING POOL, RAMSGATE.

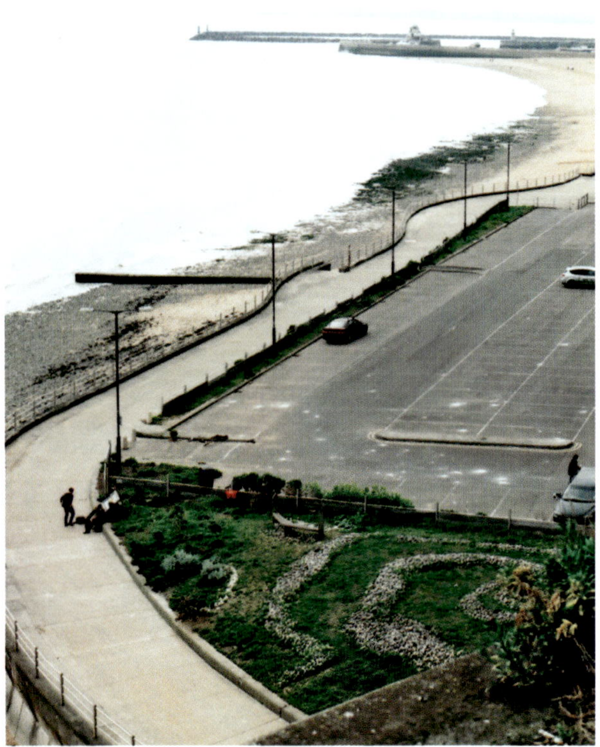

An Alternative Theory

Another popular theory concerning the demise of the Marina Bathing Pool is that the problem lay in the old design of the building, which was actually standing on stilts and at low tide you could walk underneath it. As the tide came in, the void was filled with seawater. If this is so it was a fatal design flaw. Gradually, the chalk below the structure began to erode as you would expect it to and the foundations moved, causing the pool to crack. This was patched up for years until repairs became impossible. It was also too expensive to replace the pool. Or could it just be that, as happened elsewhere, the fashion for outdoor pools and lidos just collapsed as tastes changed and it was no longer financially viable?

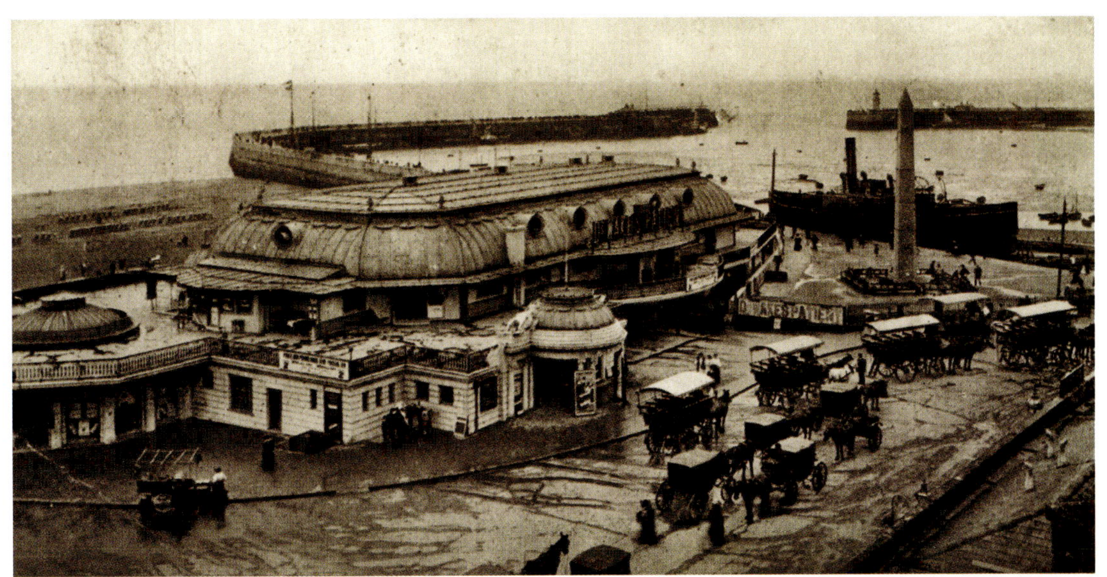

The Royal Victoria Pavilion

The Royal Victoria Pavilion or 'the Pav', as it is affectionately known, is close to the hub of Ramsgate's prosperity, and is once again becoming the focal point of Ramsgate's emerging café culture and social scene. It was opened in 1904 by Princess Louise, the second-youngest daughter of Queen Victoria, and sits next to the sands just east of the east pier of the Royal Harbour. Here we see 'the Pav' in its 1923 heyday. Since it was built it has been engulfed by the sea on several occasions, causing considerable damage and flooding. The pavilion was most recently used as a casino, but has also been a restaurant and before that, a theatre. It has recently had a major external refurbishment but its future use is still not known. I remember coming here as a young (ish!) disc jockey in the 1980s to meet the television actress Sue Pollard. At the time she was playing the part of Peggy Ollerinshaw in *Hi de Hi!* and Pleasurama, who ran a nightclub here, had decided to name it Peggy Sue's.

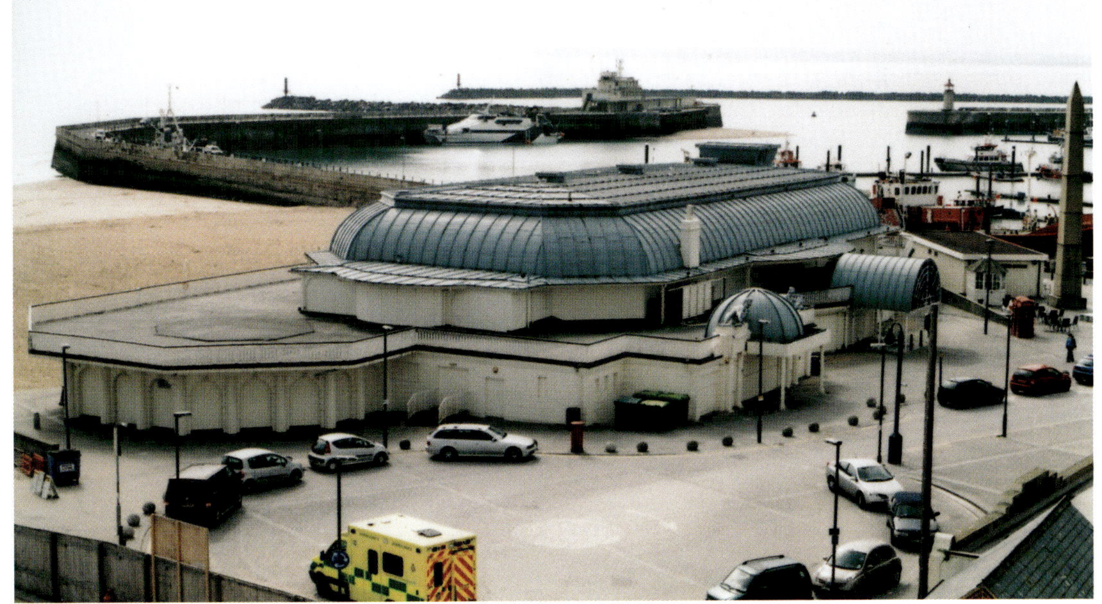

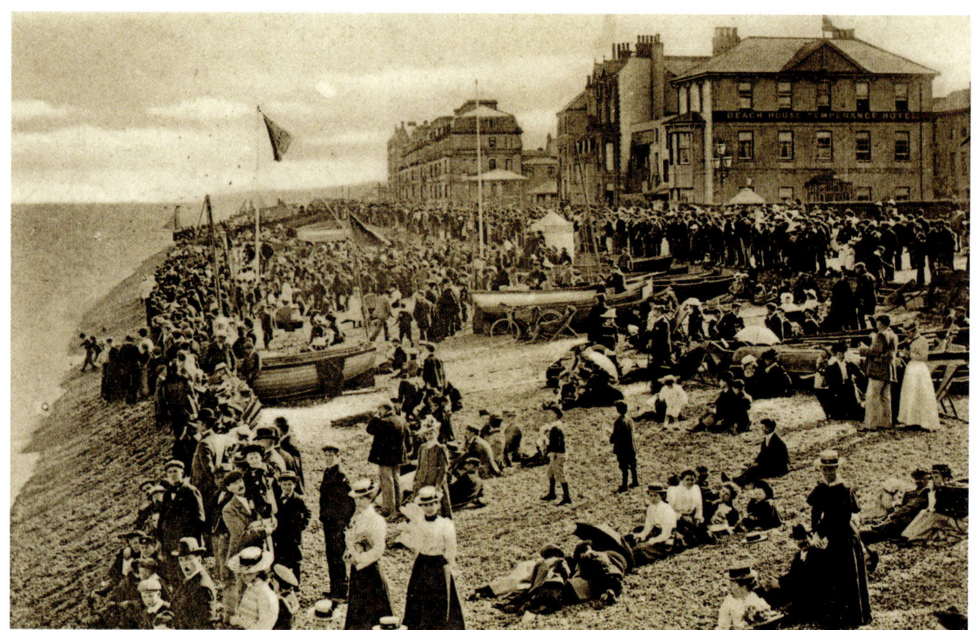

Deal

In this 1903 view of Deal beach it can be seen how popular the resort was, especially on a Regatta Day. Many believe that Deal's origins lay in its fishing industry, so it is somewhat surprising to learn this is not strictly true. Although fishing would have become a significant local occupation as Deal developed, it was the town's proximity to two natural features in the Channel that ensured its importance for many centuries. They were the notorious Goodwin Sands, which ensured a frequent need to rescue people and salvage ships and cargo, and the stretch of sea between the town and the sands known as The Downs. It offered a safe haven for shipping in bad weather and a convenient place to take on crew, passengers, pilots and supplies, a service that Deal boatmen were delighted to provide. By the early 1700s, Deal ranked as a major English naval port alongside Portsmouth, Plymouth and Chatham. Even though it lacked a harbour or docks, Deal had a dockyard on the stretch of land between the beach and today's Victoria Road. After its closure in the 1800s, the land was redeveloped for housing.

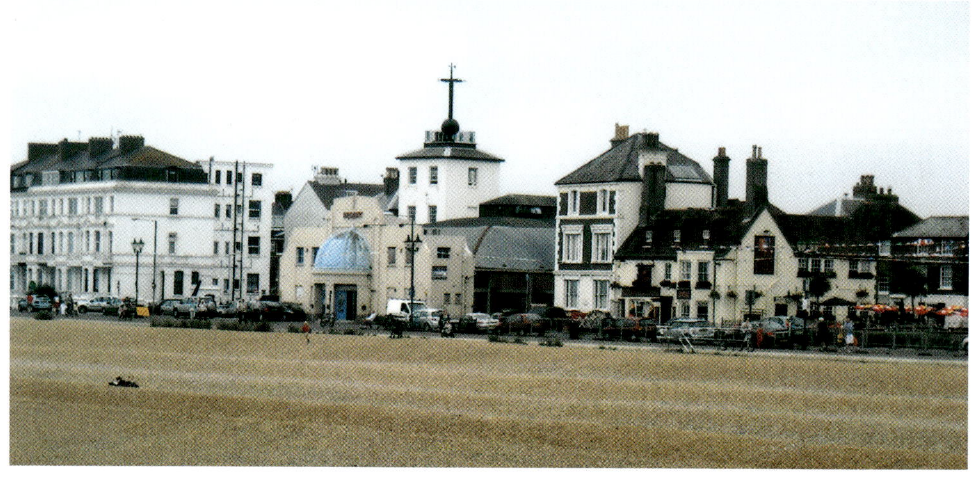

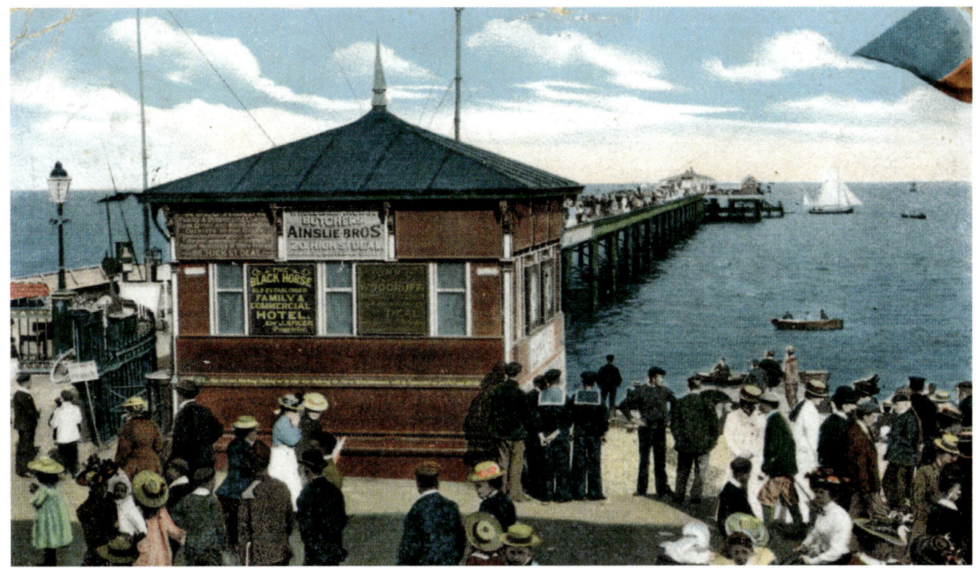

Deal Pier

Deal has the only remaining complete leisure pier in Kent. It was the third to be built here and was opened in November 1957 by HRH The Duke of Edinburgh. It is a concrete-clad steel structure, measuring about 1,026 feet, and was designed by Sir W. Halcrow and Partners at a cost of £250,000. The first pier was designed and built by Sir John Rennie in 1838. It was a wooden structure, planned to be 445 feet long, but only 250 feet was completed due to financial problems. Steamers regularly called here, but the pier decayed due to storm damage and sandworm attack. It was washed away in 1857 and its remains were sold for £50. Work on a new, 1,100-foot pier began in the spring of 1863. Designed by Eugenius Birch, pier number two opened on 8 November 1864. A reading room and salt baths were added in the 1870s and a pier-head pavilion in 1886. Ship collisions in 1873 and 1884 caused damage that was subsequently repaired, and Deal Council bought the pier for £10,000 in 1920. On 29 January 1940, the Dutch vessel *Nora*, having been beached following damage by a mine, drifted against the pier, destroying 200 feet of ironwork. A feature of this pier is a 300-centimetre-high bronze statue, *Embracing the Sea*, by sculptor John Buck, at its entrance. It was commissioned in 1998.

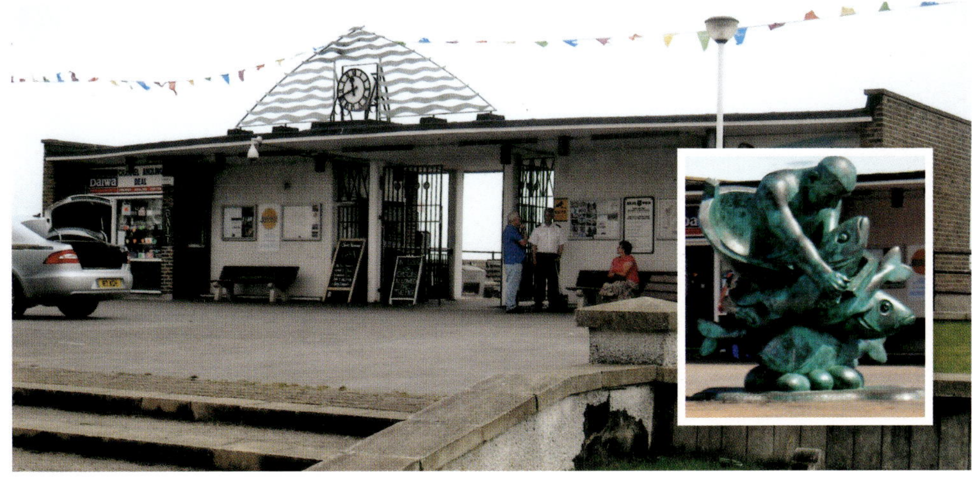

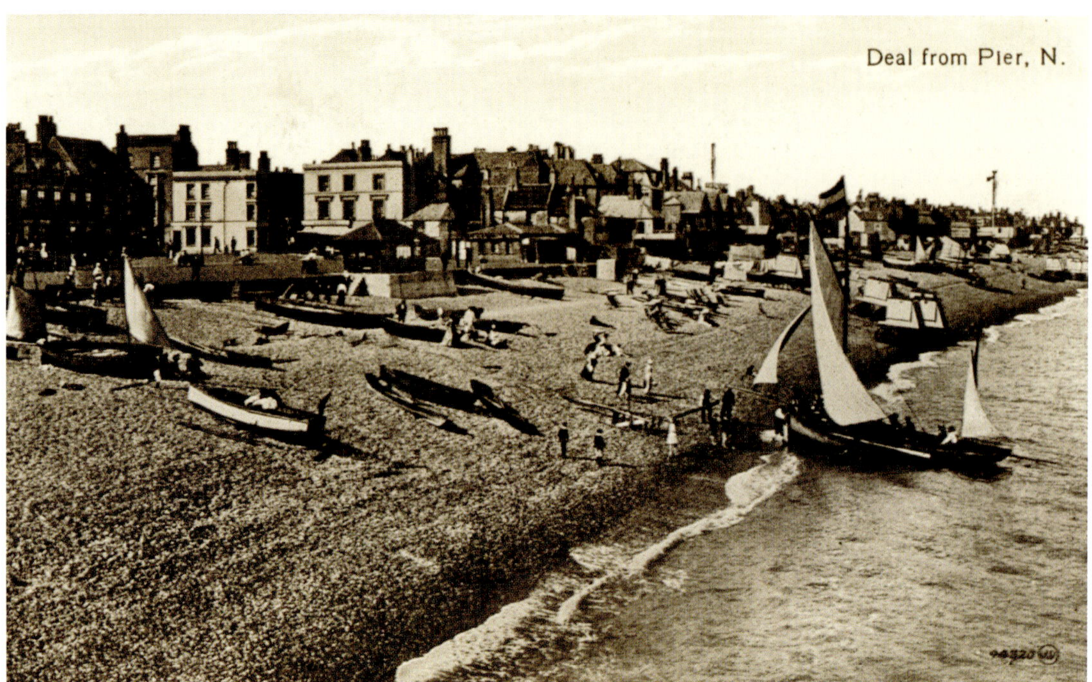

The North Beach

Looking northwards along the beach from the pier in 1918. This end of Deal's seafront has many smaller properties from the Georgian period onwards. Some are referred to as being former fishermen's cottages.

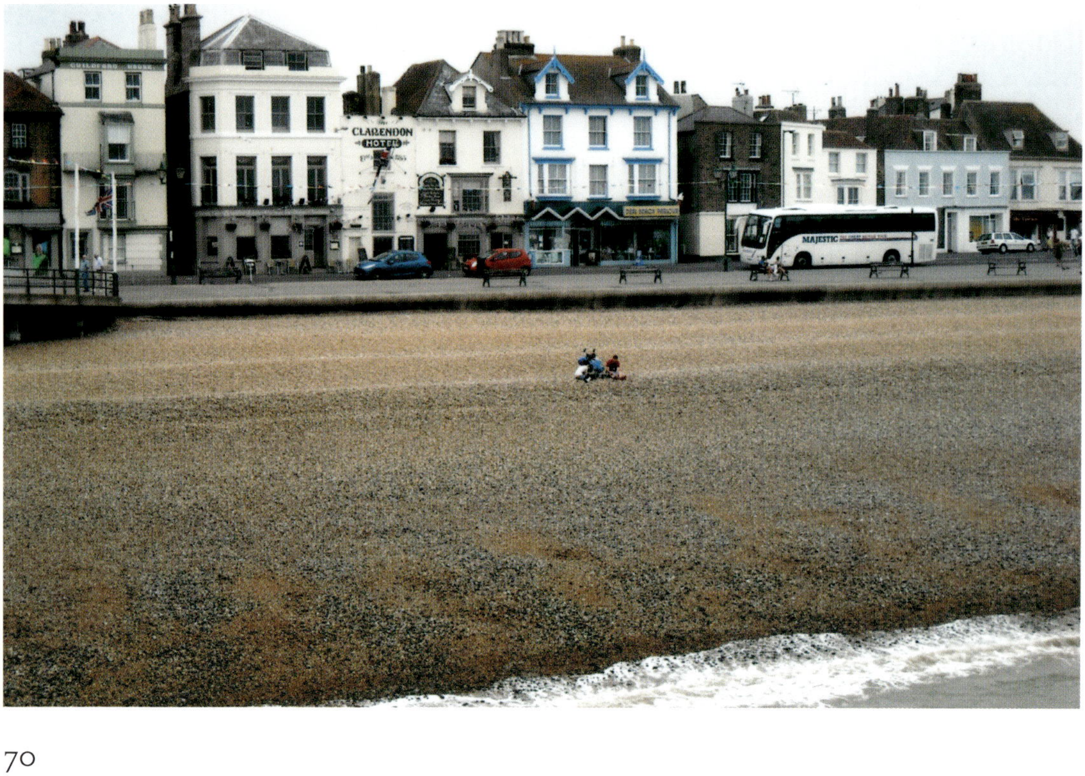

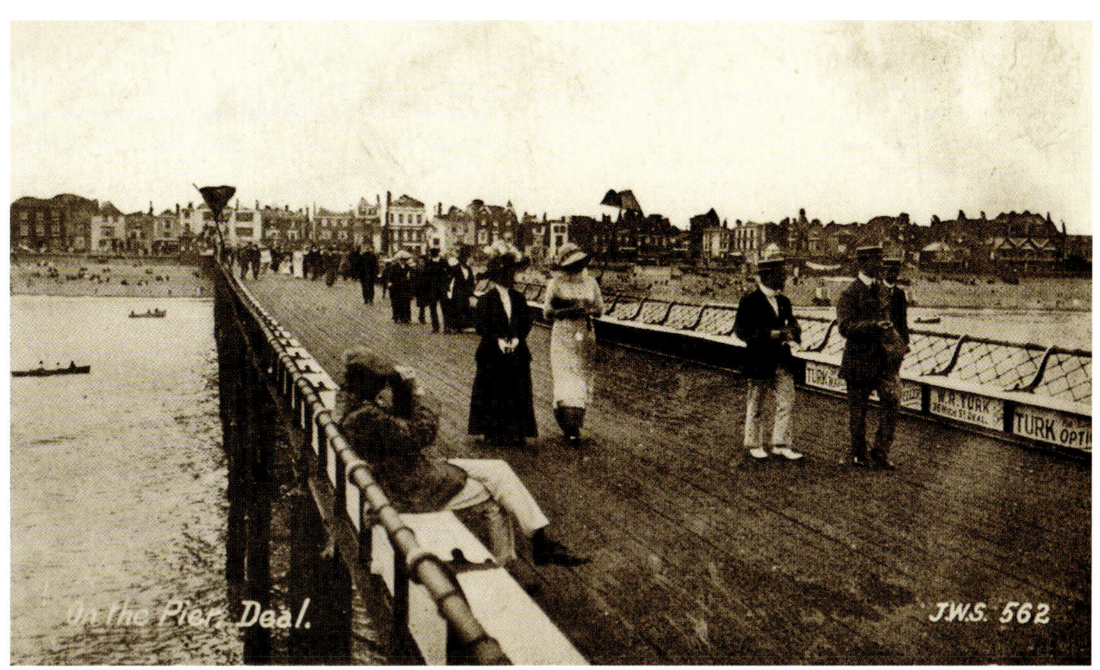

The South Beach

The seafront as seen from the pier in the early 1900s, judging by the long dresses, fashionable blazers and white flannels. This view shows just how popular promenading along the pier was. Today's view of the south beach shows the many restaurants, cafés and bars now there.

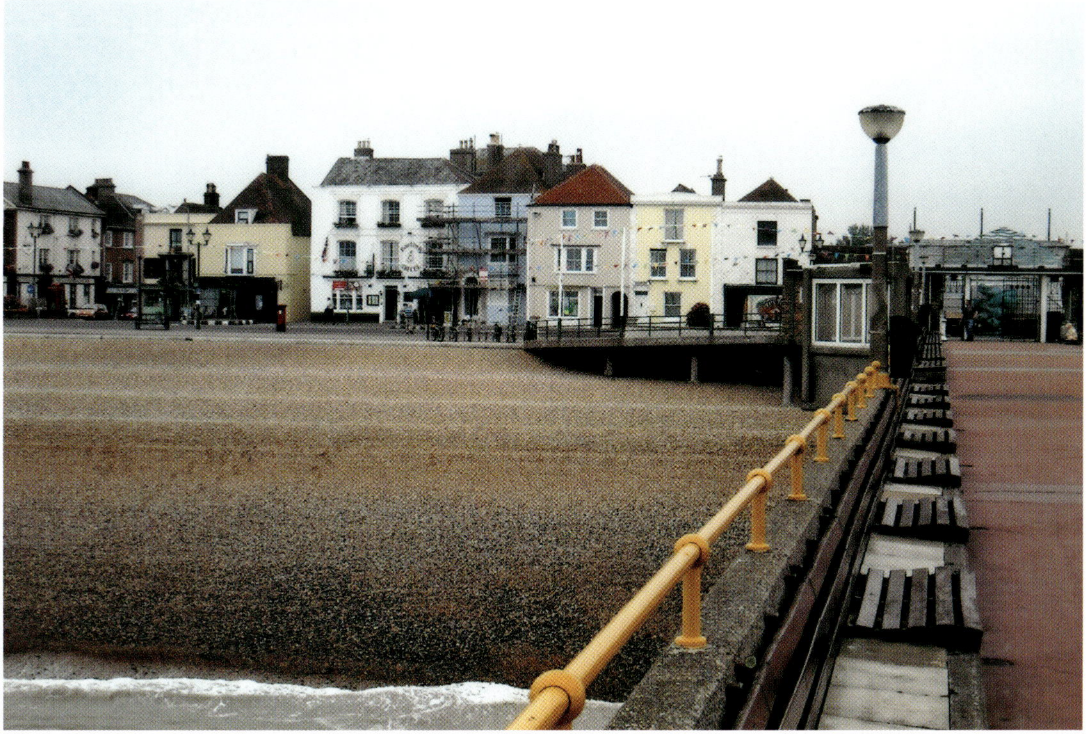

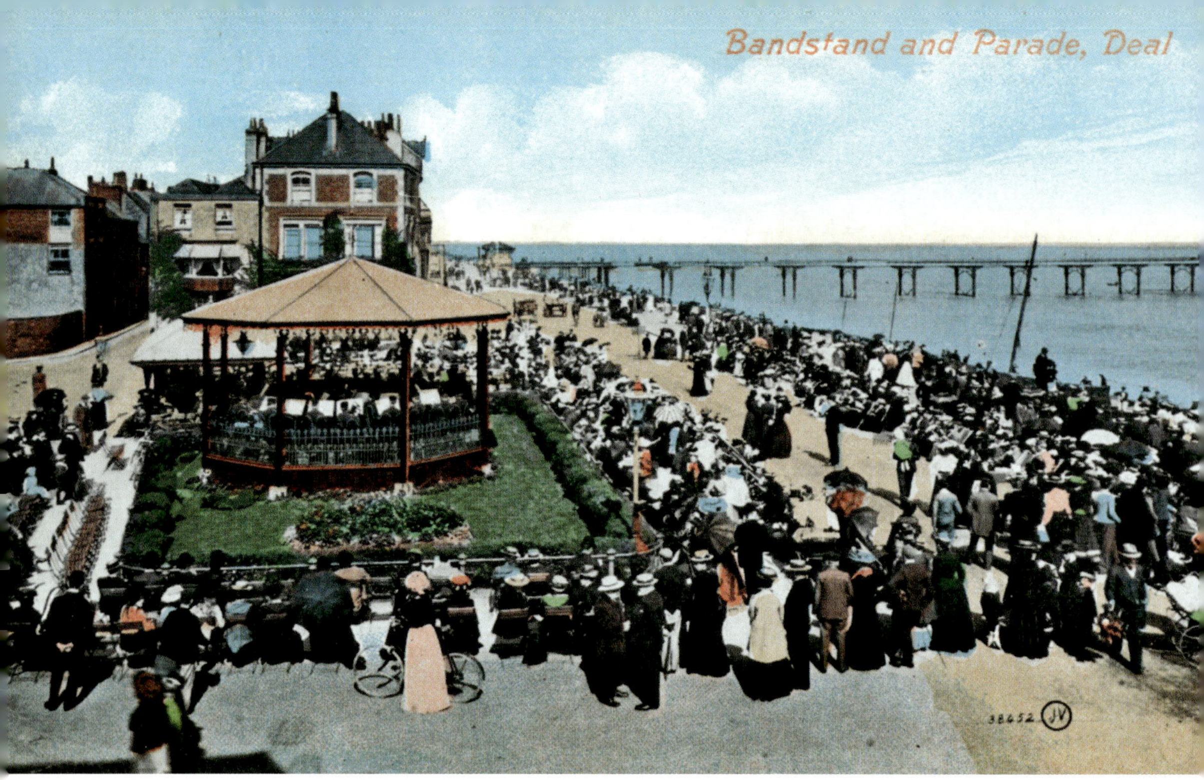

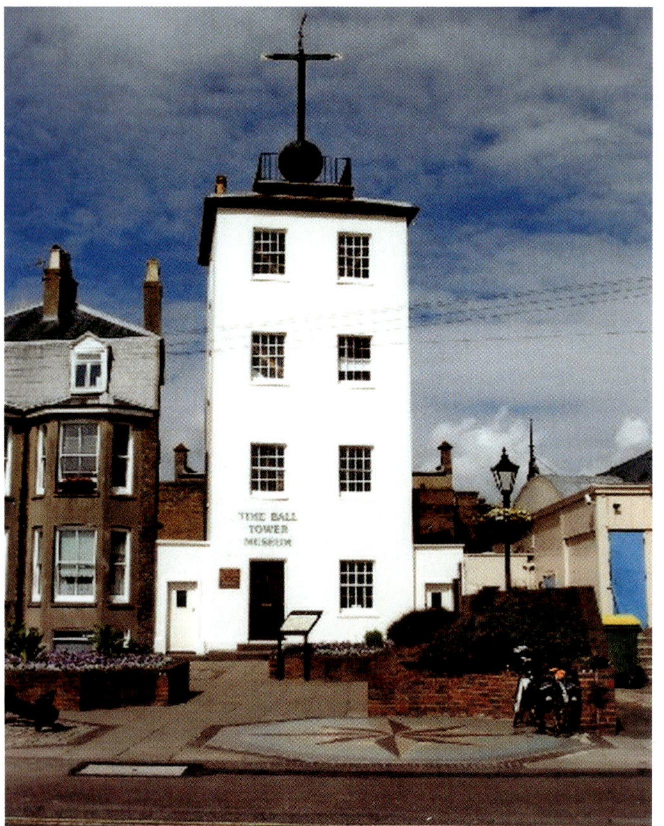

The Bandstand

For many years visitors enjoyed music performed at a bandstand that once stood in front of the Timeball Tower, as seen in this 1907 view. I have not been able to ascertain when it was demolished, but in 1992 a new Memorial Bandstand was built a mile or so along the coast on Walmer Green as a memorial to the eleven Royal Marine Bandsmen who tragically lost their lives as a result of the bombing of their barracks by the IRA on 22 September 1989. The twelve-sided structure has engraved plaques bearing the names of the eleven musicians who lost their lives.

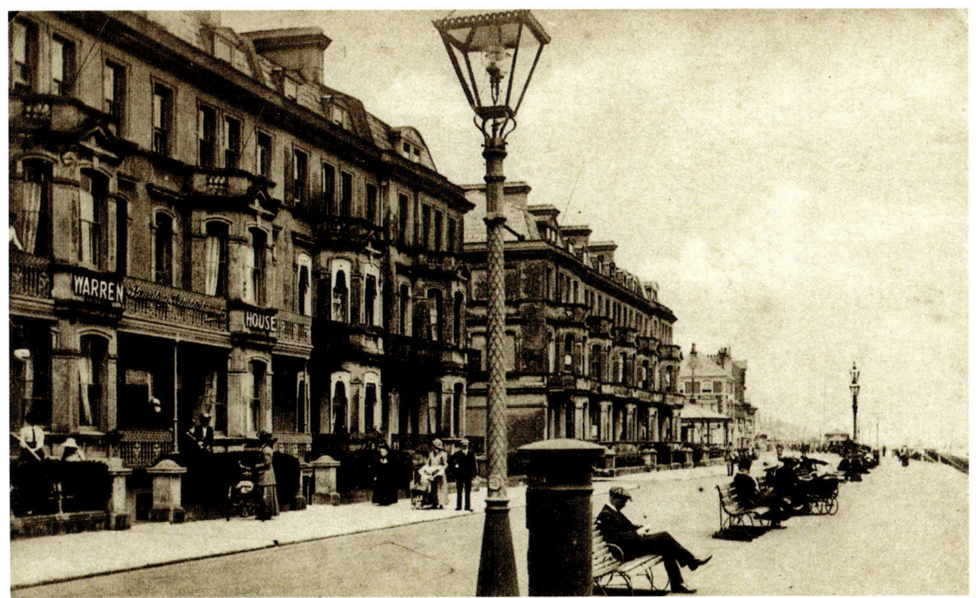

Warren House

Was this once a grand hotel or just a block of apartments? Signage on the first-floor level suggests it was a boarding establishment. Extensive research has failed to reveal exactly what this establishment was but what can be said with a degree of certainty is its location. On the original 1912 view can be seen the bandstand, just beyond the second block of properties. It is known at this time that the bandstand stood in front of the Timeball Tower. In addition there is still a post box on the promenade; not the original by any means but a post box nonetheless. The siting of post boxes is strictly regulated by the Post Office, so there being one here now means that it is more than likely that one was there in 1912. Furthermore, as this postcard was sent by someone then living at Warren House, it might well have been posted in this very post box. By 1938 this establishment had become the Deal House Hotel, whose advertisement can be seen in the town guide for that year.

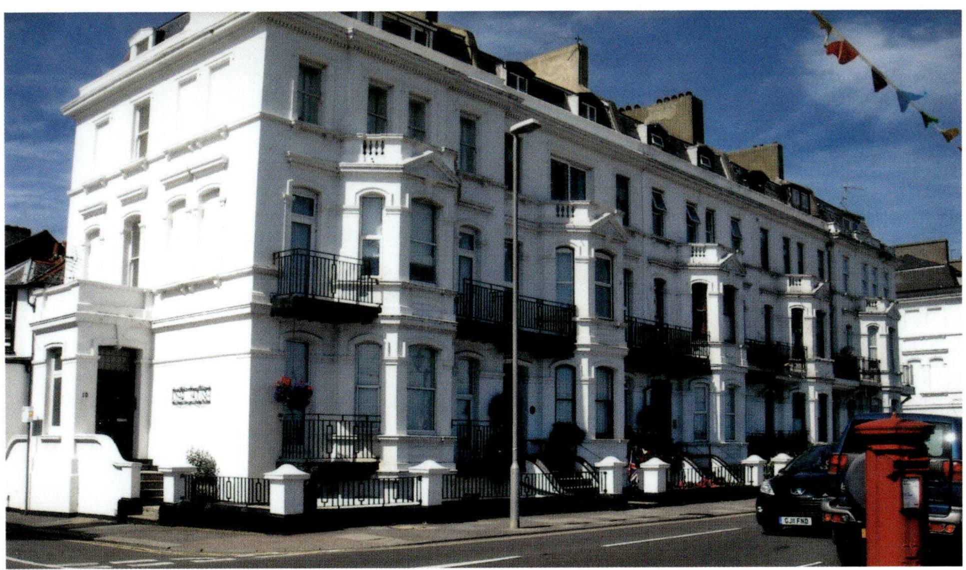

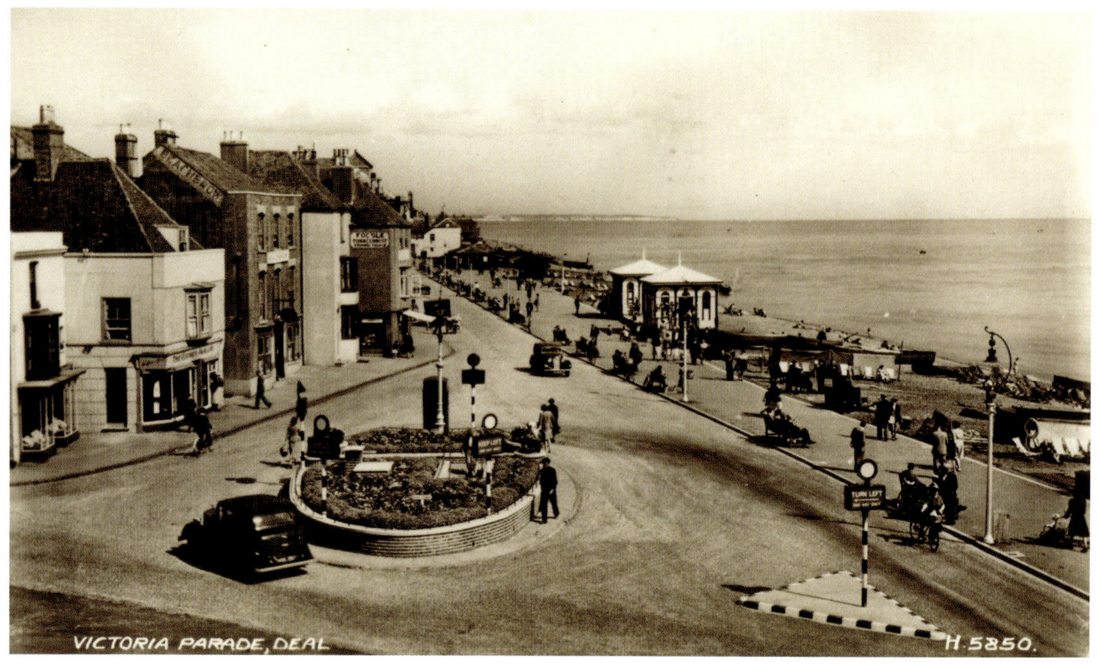

Victoria Parade

As the promenade, Victoria Parade, turns sharply right at this point to leave the seafront and skirts the castle, it can be seen how as late as the 1950s there were several small shops and other properties on this corner. More recently they have been replaced by a modern block of flats that, while offering very little architectural appeal, gives expansive sea views for the residents.

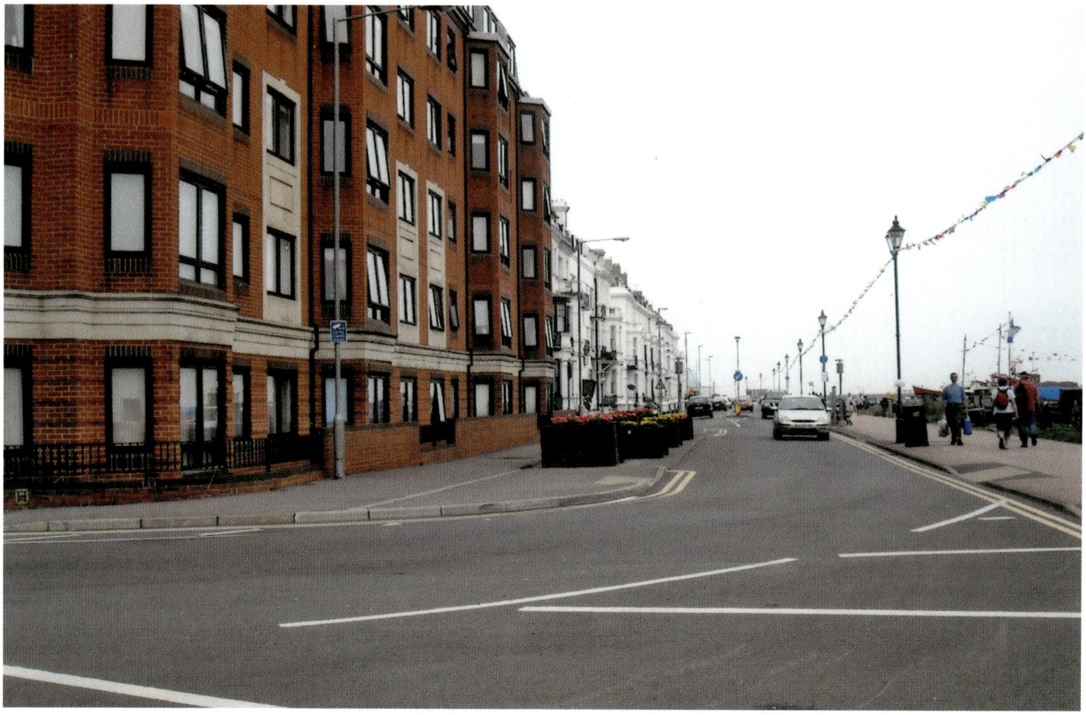

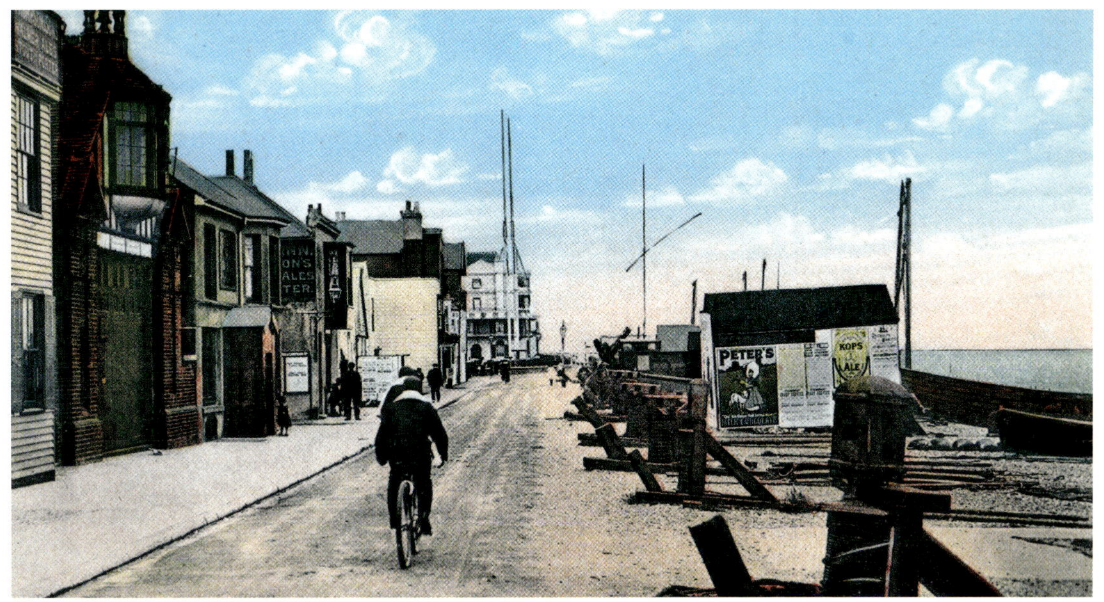

The Lifeboat Station

This undated view of Beach Street North overlooking the fishermen's beach clearly shows Deal's lifeboat house. It closed in 1932, having been in service for sixty-seven years, and in that time had five lifeboats. Today, the building serves as the headquarters of the Deal Angling Club. The main lifeboat station is now at Walmer and was built in 1856. The Walmer lifeboat station was originally a wooden boathouse but after a larger lifeboat was purchased in 1871, a new, brick building had to be put up for her. It was built in a Neo-Gothic architectural style and appears to be very similar to St Saviour's church, which stands directly opposite along the Strand.

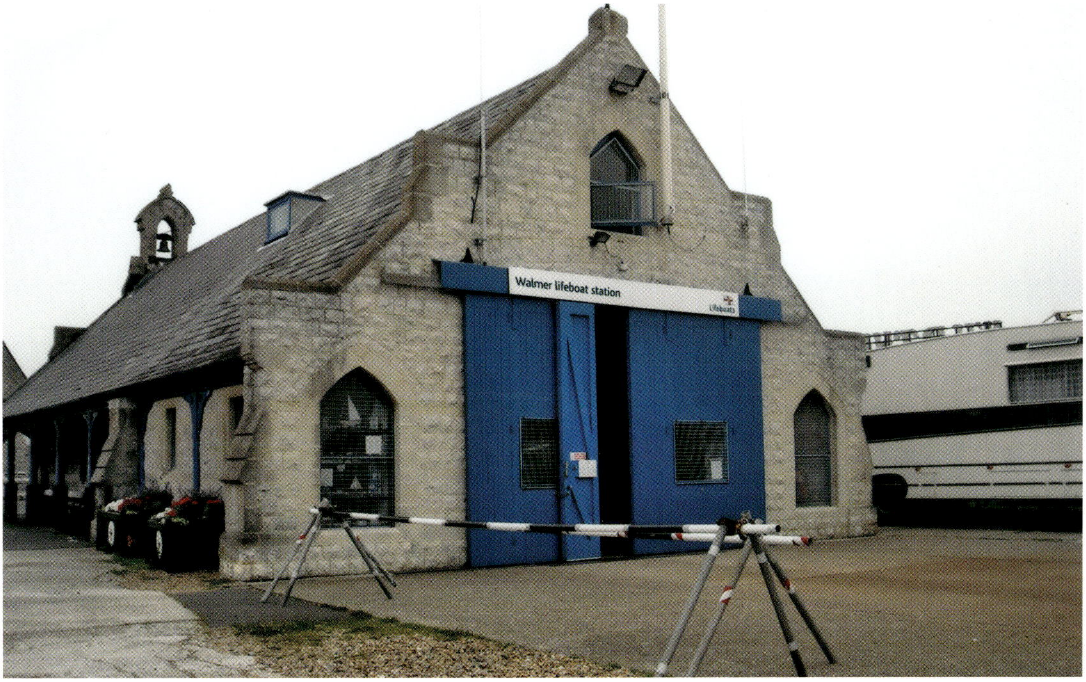

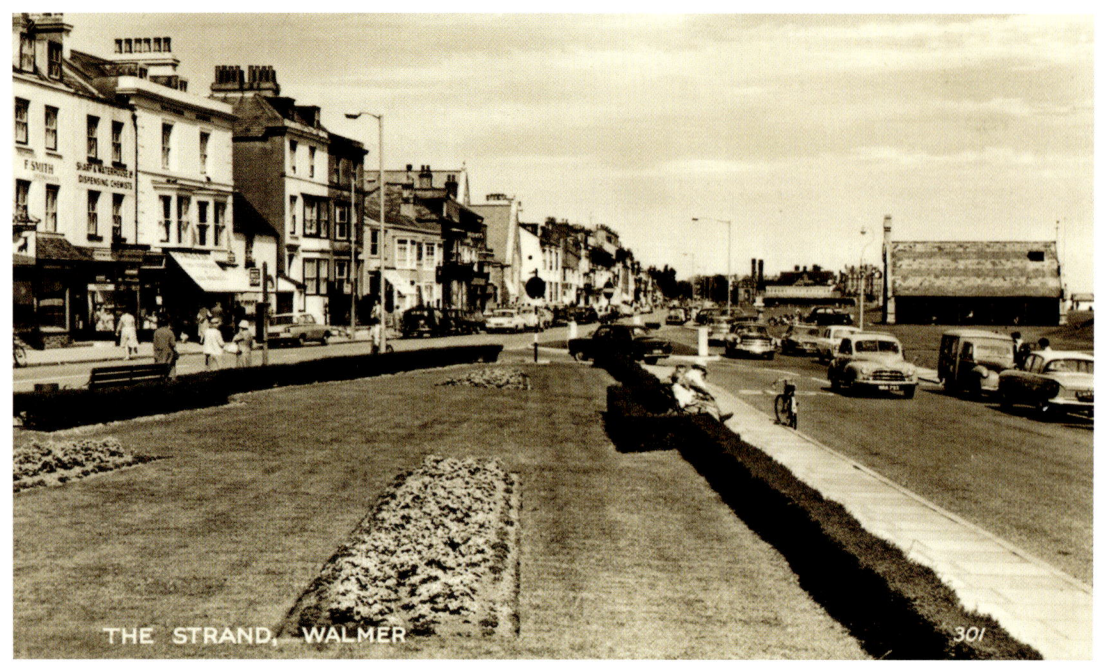

The Strand, Walmer

Many seaside resorts have an 'up-market end' and Deal is no exception. Walmer has a through-road, but it was once a pleasant drive with a grass and floral displays central reservation. Today, most of it has gone, replaced by a wider main road. It is here that you will find Deal's memorial bandstand, amidst a wide grassy thoroughfare.

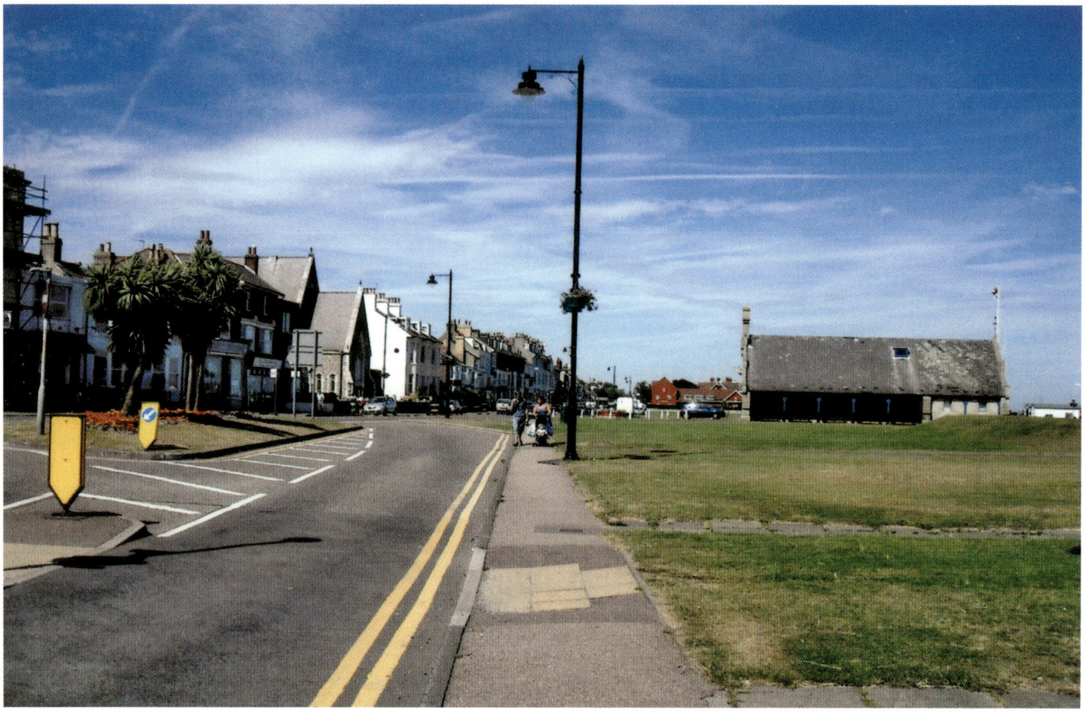

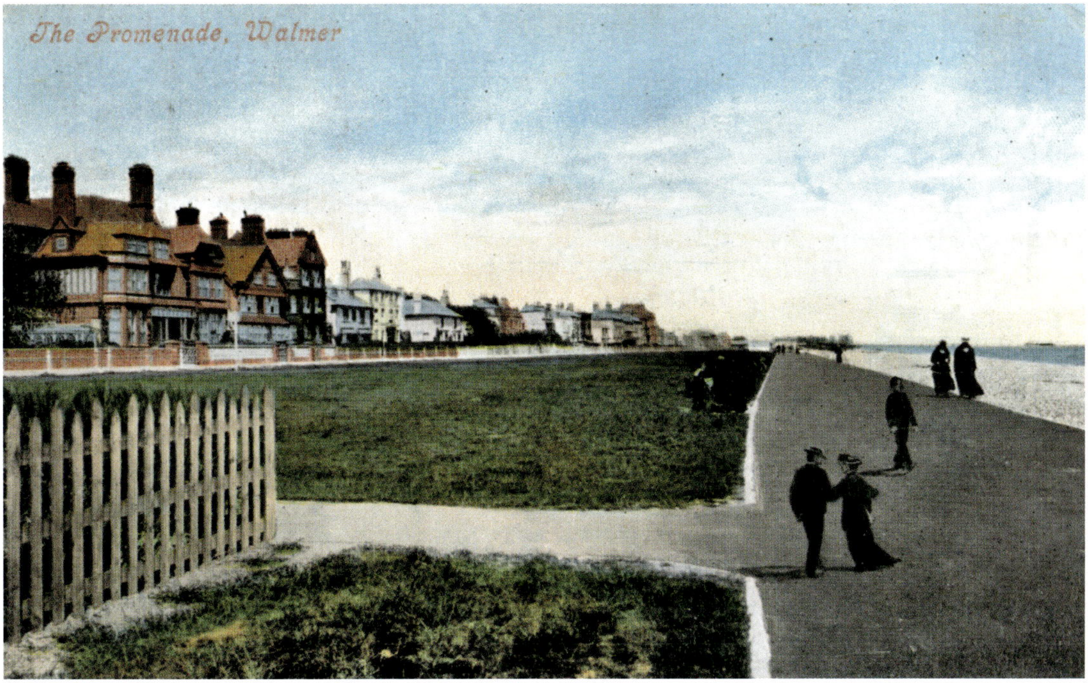

Walmer Promenade

A century or more separates these two views of Walmer Promenade but apart from the difference in dress, it can be seen that little has changed. Many of the old Edwardian villas are still standing, one of which was owned by Joseph Lister, the pioneer of antiseptic surgery.

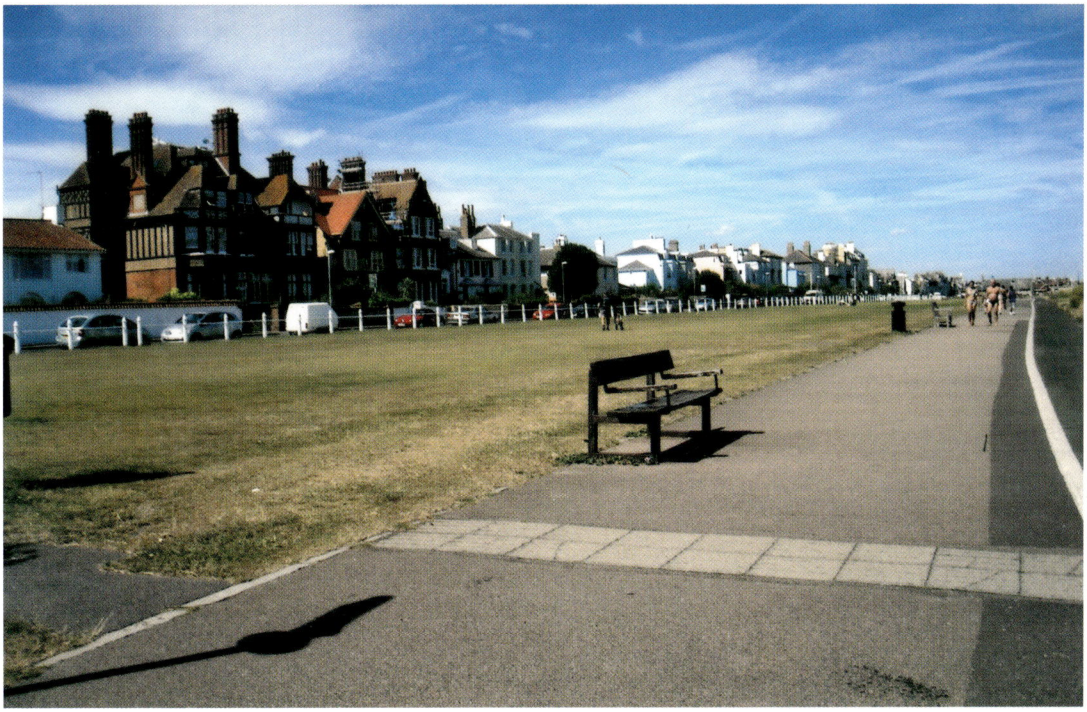

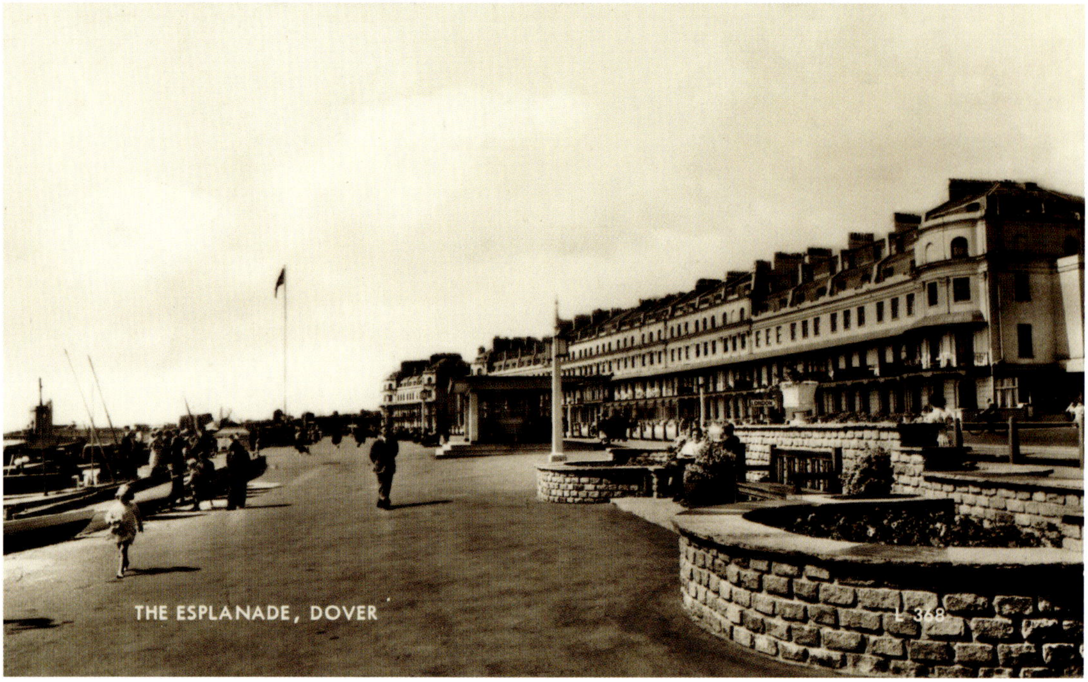

Dover Esplanade

Although not strictly a seaside resort, more a ferry port, Dover has a well-laid-out promenade and a small beach overlooking the harbour. In this early view it can be seen how small boats were pulled right up the beach, almost onto the promenade itself.

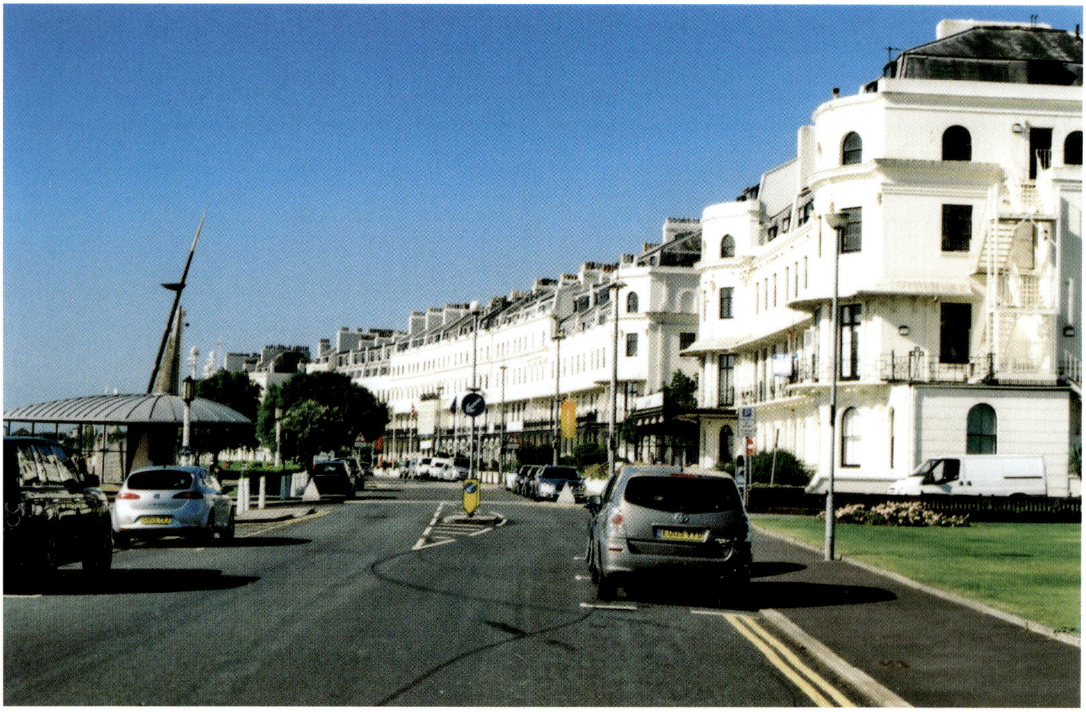

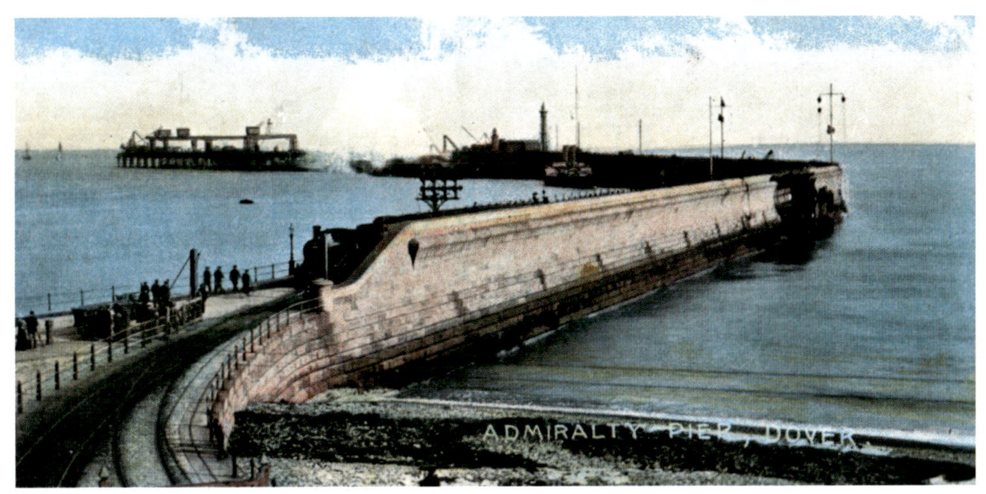

Admiralty Pier, Dover

Building works on the Admiralty Pier were started on 2 April 1848 as the first stage in a proposed harbour for the Royal Navy. The first section of pier was 800 feet long and was completed in 1854, when another contract was awarded for a further 1,000 feet. Work on this extension began straight away and should have been completed by November 1864, but a decision had not been reached on how to terminate the pier. It was not until 1871 that it was finally decided to add a further 300 feet, terminating in a substantial pier-head. The work was completed in 1875 and in the late 1870s it was decided to build the Pier Turret on the pier-head for defence. From 1851 cross-channel steam ships used the pier regularly and in 1860 the South Eastern Railway started running its trains along the pier to connect with the railway-owned steamers, and then in 1864 its rival, the London, Chatham and Dover Railway, did the same. In 1899 work started on extending the pier again as part of the new Admiralty Harbour and in 1900 the 2,000-foot extension was completed, thus bringing the total length of the pier to 4,140 feet. In 1909 work started at the landward end of the pier to reclaim land for the building of the new Dover Marine railway station – the remains of which can be seen in the modern-day photo – which was completed in 1921. In 1981 a special berth was constructed for the Jetfoil service to Ostend, and a new train ferry berth was opened in 1988 to replace the old Train Ferry Dock. In 1994–96 the old Dover Marine station was converted into the new Cruise Liner Terminal, and in 1998–2000 a second cruise terminal was constructed, involving the widening of the pier extension.

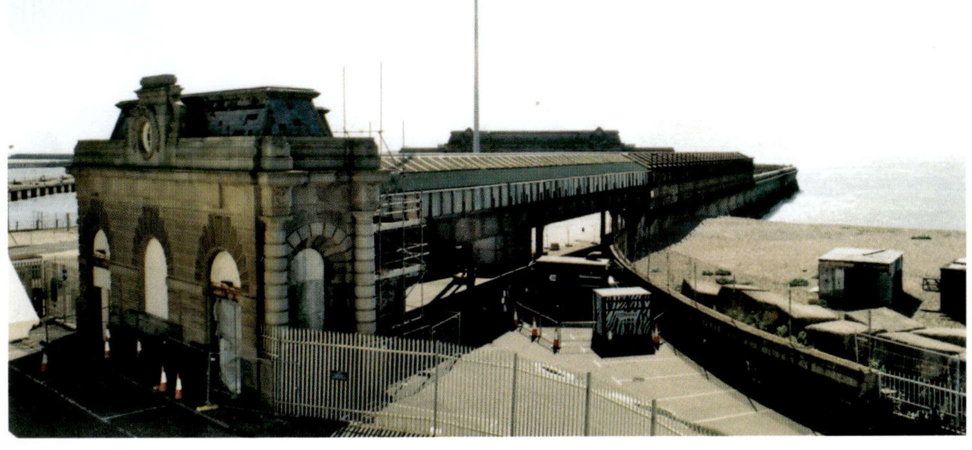

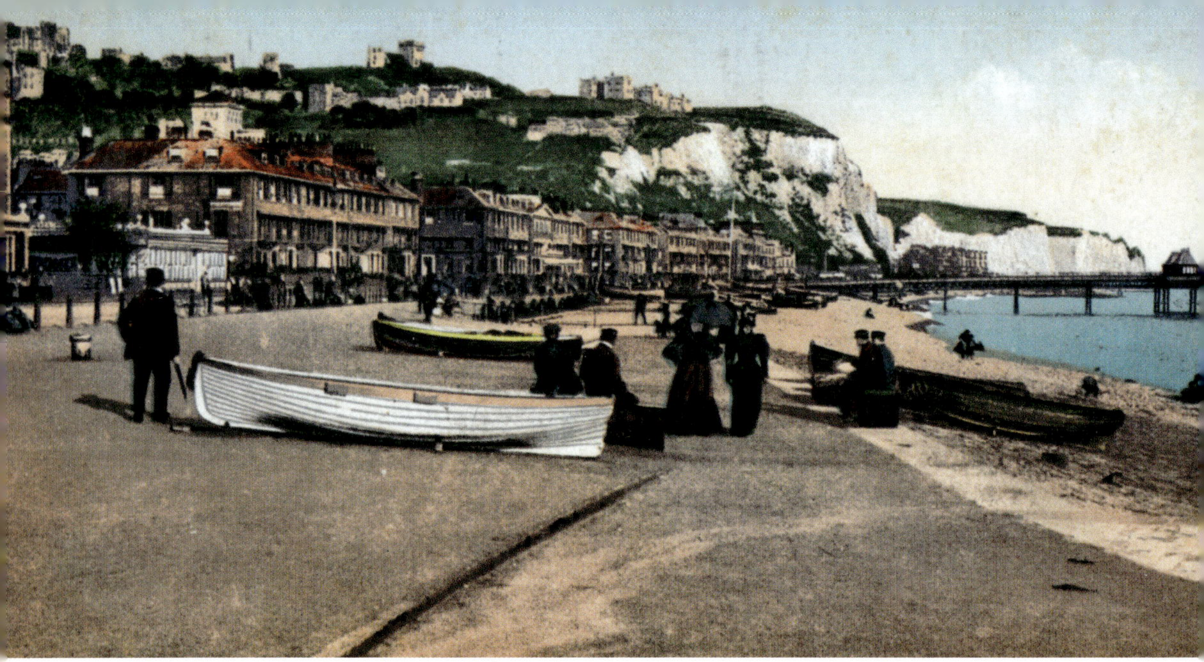

Dover's Beach

With the white cliffs rising up behind you and the sea vista of a busy harbour, Dover's beach is not the best of beaches to relax on. No doubt when this Edwardian view was taken the harbour would not have been as busy as it now is, but it is littered nevertheless with small fishing boats. Even to this day the cliff-top castle dominates the town.

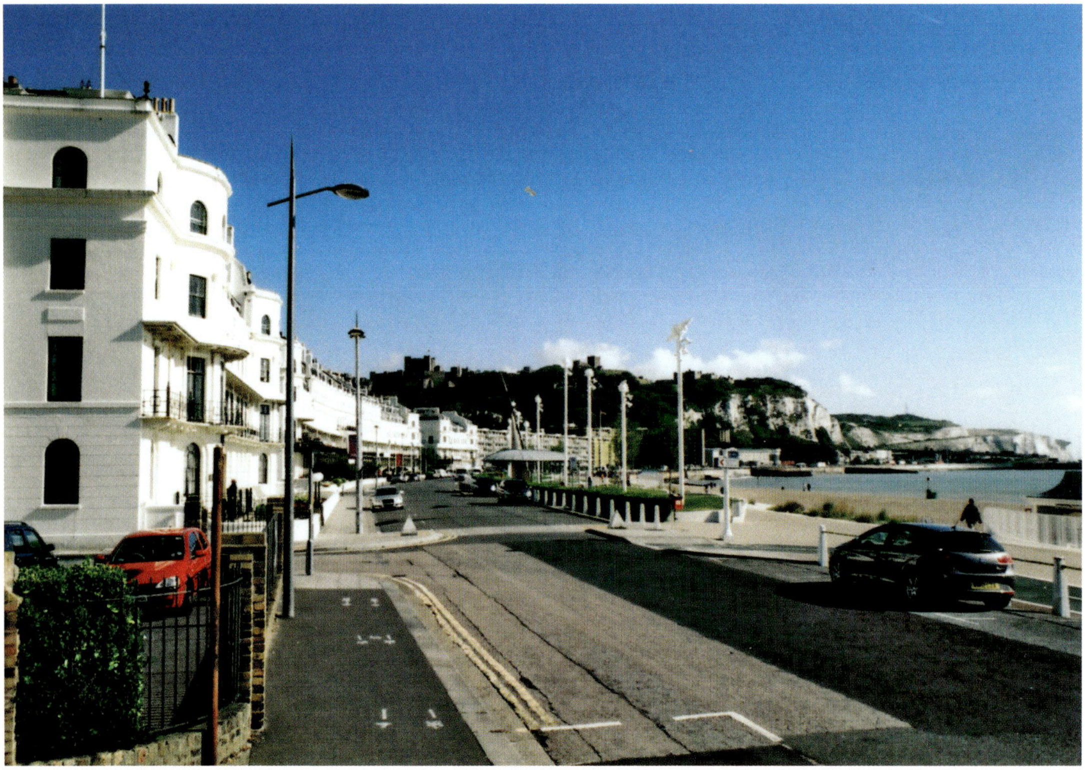

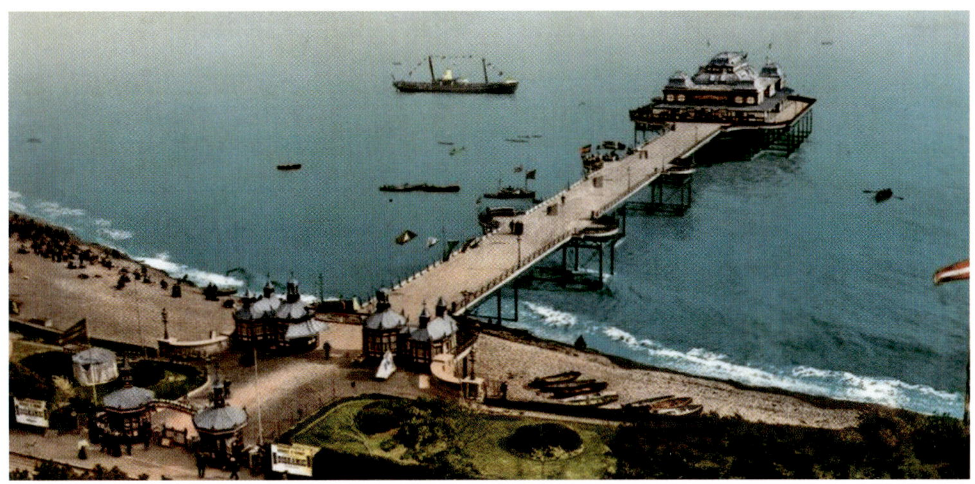

Folkestone Pier

It was not until I began researching the history of Folkestone that I discovered it once had a pier. It was the brainchild of local estate agent George Bramston Trent, who formed the Folkestone Pier and Lift Co. in the early 1880s, originally proposing an 800-foot pier. The foundation stone was laid on 7 May 1887 and the Victoria Pier, named in honour of the 1887 Golden Jubilee, opened on 21 July 1888; it was designed by M. N. Ridley. The finished length was 683 feet and included a 700-seat pavilion, which was leased out to theatrical companies, who provided suitable high-brow entertainments for Folkestone's largely aristocratic clientèle. A floating landing stage was added in 1890 but was rarely used. The pier was popular but not profitable until 1898. In 1903, Keith Prowse & Co. Ltd took over the lease on the pavilion and introduced variety performers, including Lily Langtry and Dan Leno. In 1907 the lease on the pier transferred to local businessmen Robert and Lloyd Forsyth, who abandoned expensive entertainers in favour of more profitable attractions such as wrestling, a cinema and beauty contests. In 1910 the Olympia roller-skating rink was added on the shore to the west of the pier. The pier was closed and sectioned in 1940 for defence purposes. A temporary bridge was installed in 1943, but a fire on Whit Sunday wrecked the pavilion and badly damaged the seaward end of the pier. The remains were demolished in 1954. Today, only the abutment to the pier and a small section of an iron supporting column survive to show where it once stood.

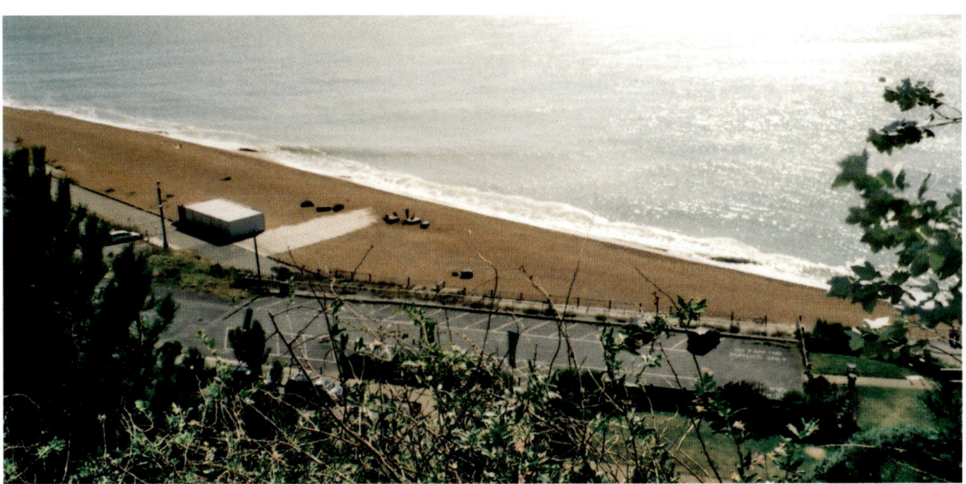

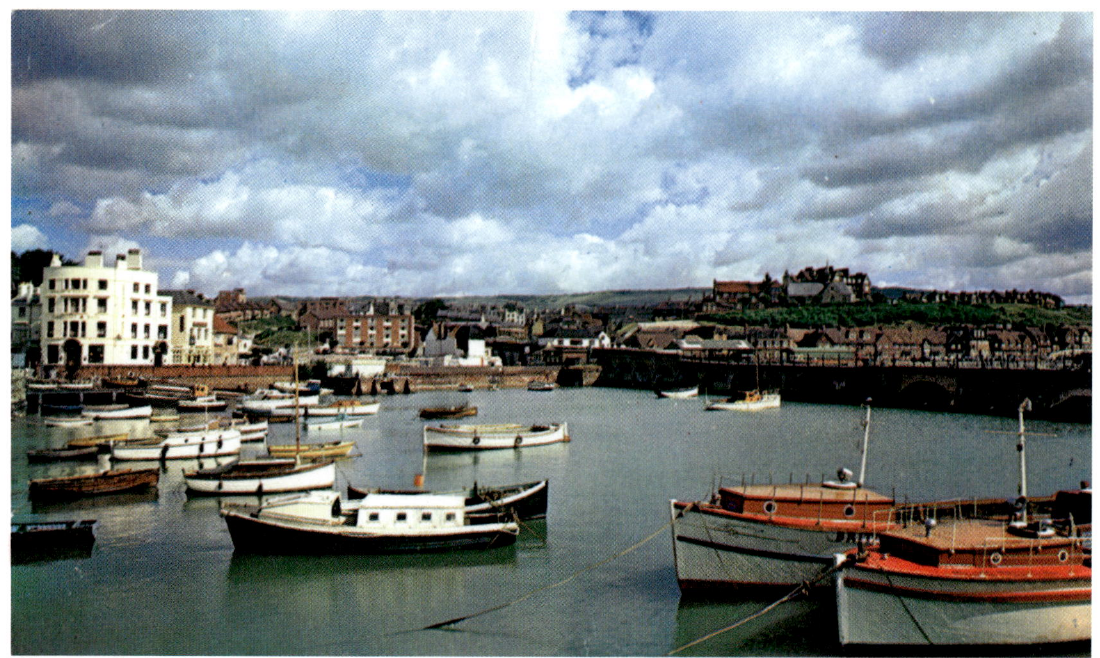

Folkestone Harbour

Folkestone Harbour was built in stages between the 1840s and the early 1900s by the renowned engineer Thomas Telford, around the old fishing port of Folkestone. Over the years, passenger and freight ferries, as well as cargo ships, have operated from this port to various destinations, the best-remembered link being that between Folkestone and Boulogne. By the late 1990s, ferry operators Sea Containers Ltd, who then also owned Folkestone Harbour, decided that ferry operations had become unsustainable and the last ferry link was closed down in 2000.

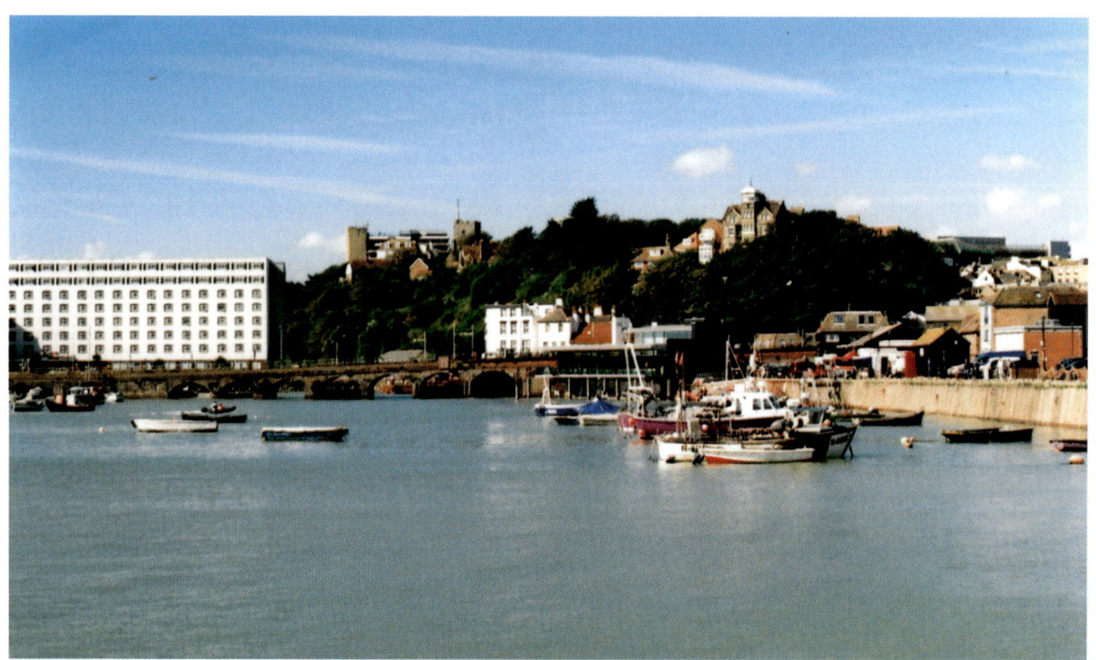

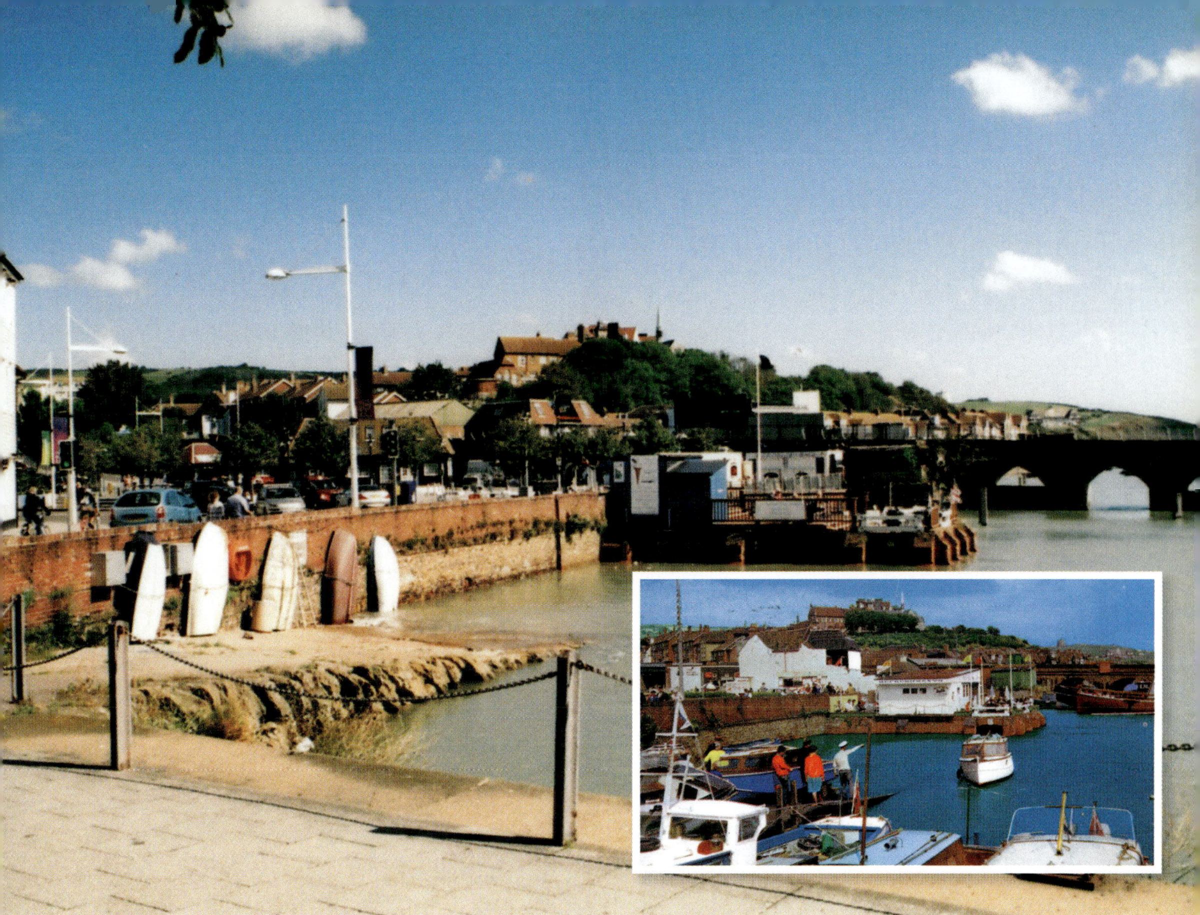

Folkestone Harbour

Folkestone has long been recognised for its cross-channel ferry service to Boulogne, and 1 August 1993 marked the 150th anniversary of the opening of this route to regular traffic, a date fixed by the coming of the railway to the port. In June 1847, *The Times* noted that with the opening of the Boulogne and Amiens railway to Abbeville, it was now possible to reach Paris from London in fourteen hours, the Folkestone–Boulogne crossing taking one hour and forty-five minutes. London to Paris (via Folkestone–Boulogne) was further reduced to eight hours in 1884 and a through-journey time of seven hours and thirty minutes became possible in 1891. The line from London had arrived in June 1843 but as the nineteen arches of the Foord Viaduct were still under construction, a hastily built terminus was erected near the site of the present Folkestone Central station. It was not for another six months that the viaduct was completed and trains were able to use Folkestone station at its eastern end. By the time the railway arrived, the 19-acre site was both neglected and badly silted. Not only did the railway reach the harbour at right-angles, but in order to allow trains a level stretch in which to stop, the Railway Pier was built dividing the existing harbour into two, thus creating the Inner and Outer Harbours. At that time there was no swing-bridge allowing trains access to steamer berths on the harbour's south side and it was not until six years later that this was added, allowing passenger boat trains to operate for the first time. By further extending and widening the New Pier between 1897 and 1904, Folkestone Harbour, as we know it today, was completed. The final passenger steamer, the former Channel Islands vessel *Caesarea*, completed her last crossing between Folkestone and Boulogne in October 1980.

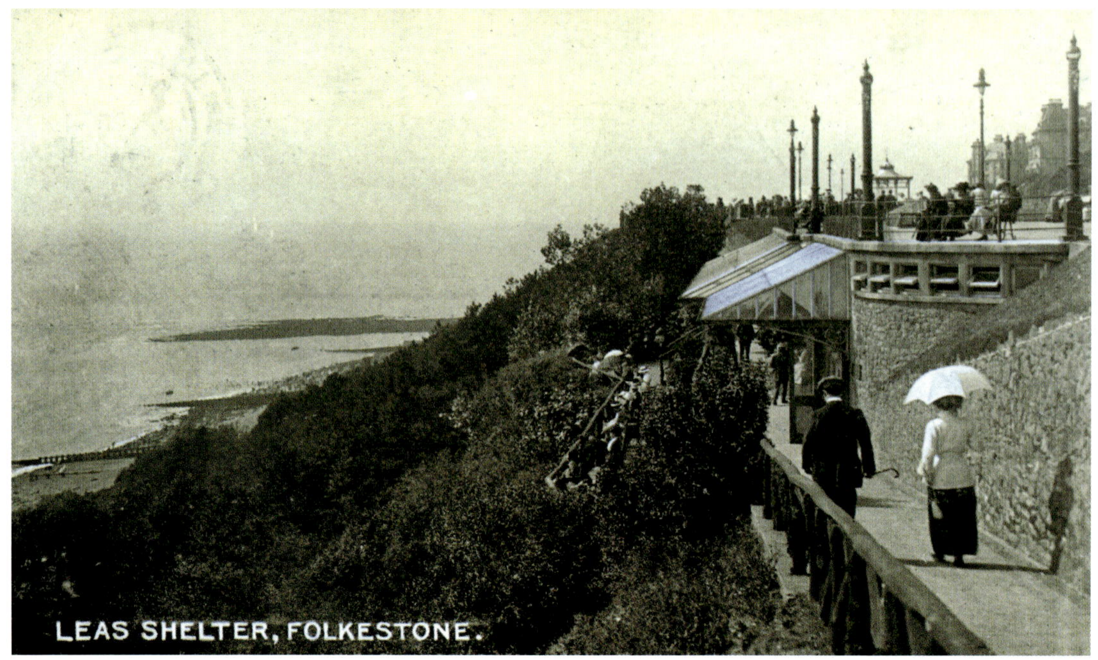

The Leas

This footpath took visitors down to the beach from the cliff top and hotels above it. The shelter was somewhere to sit and admire the expansive sea views. This 1920 view shows it to be a pleasant stroll but today, although the footpath is still there, it is buried in foliage, which is why we had to take today's view from the top of the cliff. In both views the almost obligatory bandstand can be seen in the distance.

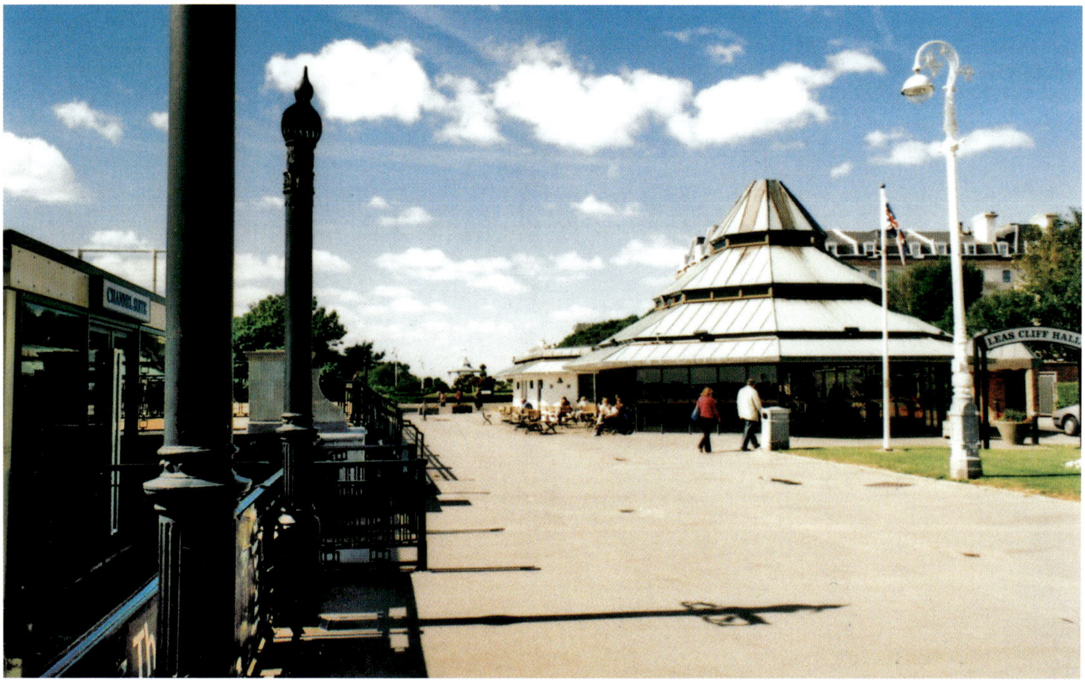

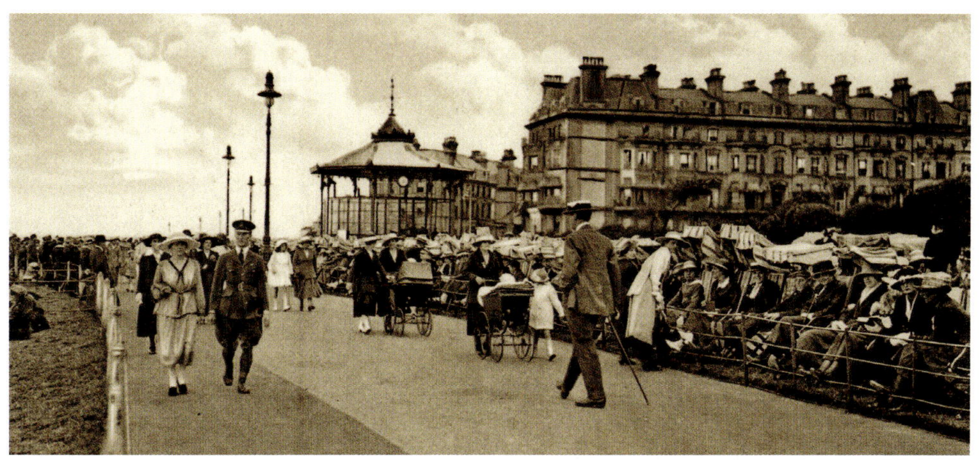

The Bandstand

Built during the Victorian era, Folkestone's beautifully restored iron bandstand can be found on the Leas Promenade overlooking the sea from the cliff top. Folkestone has been a resort for the wealthy since Victorian times, with grand villas and hotels being built along The Leas to take advantage of the magnificent sea views. Promenading on The Leas and taking the sea air was a major pastime, but no Victorian resort would have been complete without music, and Folkestone was no exception with plenty of bands, both private and military, visiting to 'give the promenaders much pleasure'. It's hard to believe now but it took over forty years to agree on funding to build the bandstand. Many people objected to the idea, complaining that it would ruin the views of the sea. The Borough Council eventually decided to levy a one-penny rate to raise funds; this caused much anger, because working-class residents were barred from The Leas and so would never be able to enjoy the thing that they had helped to pay for. Despite the complaints, building work started in 1894 with the cast-iron structure being made at Elmbank Foundry in Glasgow and erected by builders Allen & Co. It opened in 1895 and has been regularly used for more than a century. After such a long time in its highly exposed position, by 2006 the bandstand had deteriorated substantially. The idea that such a popular landmark could be lost was unthinkable, and with funding from Shepway District Council it has now been totally rebuilt and will hopefully continue entertaining promenaders for another 100 years.

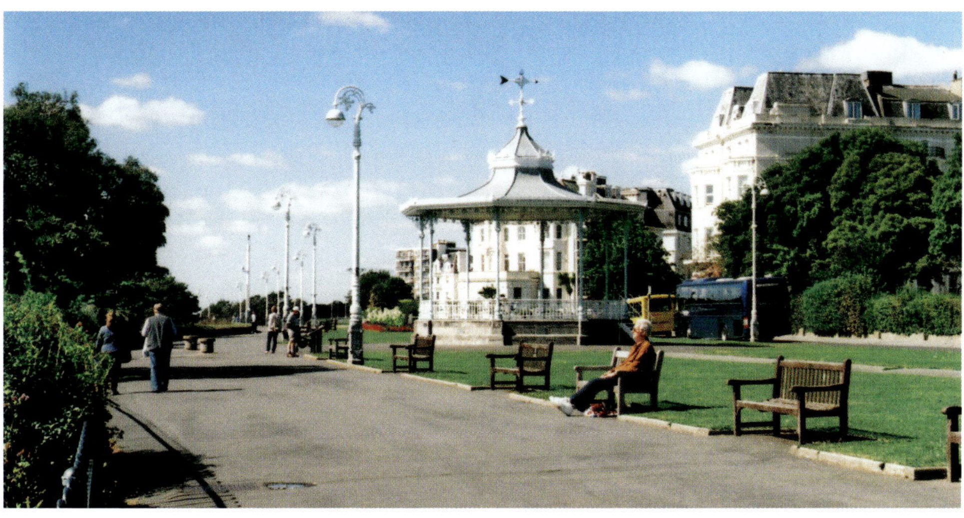

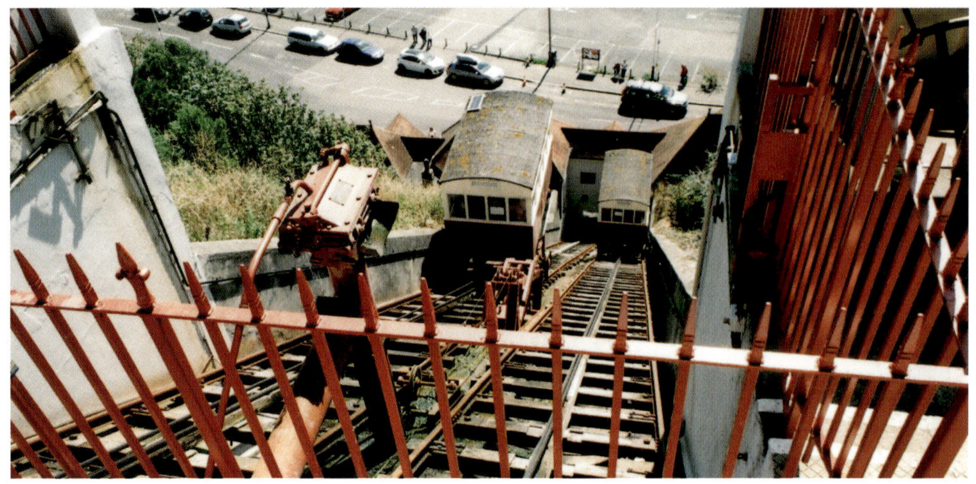

The Cliff Lift

The Folkestone cliff lift is a water-balanced funicular, opened in 1885, and for the most part remains as built. It was more successful than any other lift of its kind, so much so that following the opening of the Victoria Pier in 1888, a second lift was built alongside, and then a third and fourth further along the Leas Cliff. The original lift is still in full working order and is in daily operation during the summer months. It is estimated it has carried about 50,000,000 passengers, which is not only a testament to its original quality but also makes the carriages possibly the most used carriages ever built. Where else could you sit where several million have sat before you, overlooking the English Channel with a panorama of the busiest shipping lanes in the world and the French coast as a backdrop on clear days? The lift's principle is simple: there are two cars connected by a rope over a wheel at the top. When the brake is released, the top car's tank is filled with sufficient water to make it heavier than the bottom car and the heavier car descends, pulling the other car up the other track. When the heavier car reaches the bottom, the water is released into the bottom reservoir and is pumped back to the top reservoir, ready for re-use. The lift was originally privately owned, but was taken over by the local council in 1967. However, due to financial constraints the last remaining lift ceased operations as it was about to enter its 125th year. The council surrendered its lease to the freeholder, the Folkestone Estate, in 2009, who refurbished the original lift and its integral infrastructure, and let it to The Leas Lift CIC, which operates it throughout the year.

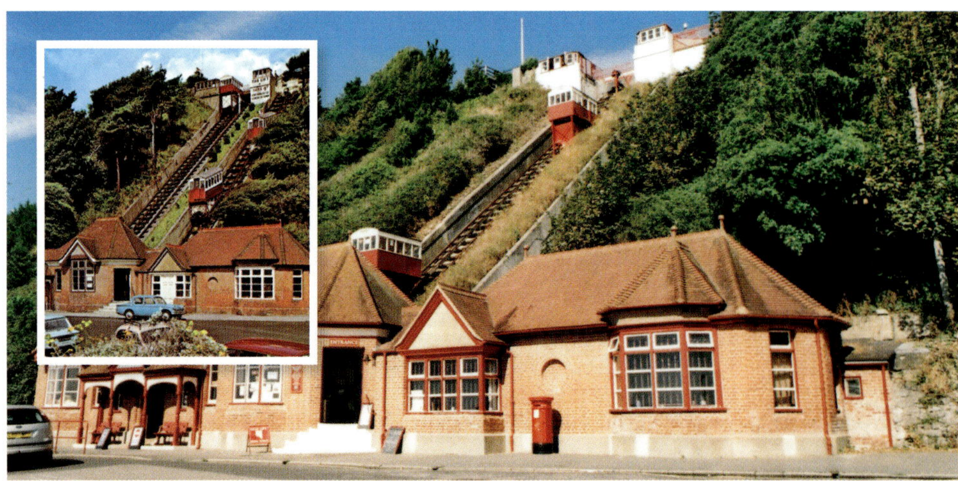

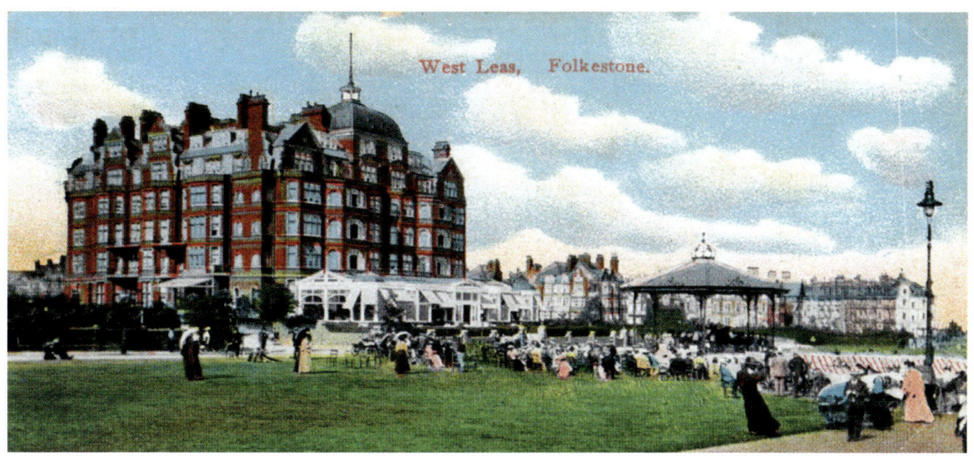

The Grand Hotel

Typical of the many grand Victorian hotels along The Leas is The Grand. When built just over 100 years ago, it incorporated many novel features which have since become the norm for all hotels. It soon became one of Folkestone's most fashionable and prosperous hotels. The Metropole, which stands immediately next door, had just been constructed, and a local builder who had been disappointed not to secure the building contract was determined to build a rival establishment that was better in every way. That builder, Daniel Baker, was at the forefront of innovative design; he had already developed the use of cavity wall ties, and went one better with The Grand by installing waterproof cavity wall insulation. He used a steel frame, one of the first, to give large, clear spans to the main reception rooms, and, said to be a world-first, infilled it with reinforced concrete. He also used suspended ceilings for improved soundproofing. Not only was he innovative, but he was also able to utilise new techniques to excellent effect. The steel frame allowed his formative use of curtain walling, resulting in the windows covering almost the entire width of the elevations to make the most of the sunny location and the fabulous views. A by-product of the concrete floors was one of the first examples of wall-to-wall carpeting. The building was constructed as gentlemen's residential chambers, and immediately established a reputation as the place to be and be seen. The king, Edward VII, became a frequent visitor, so much so that the locals would wander along The Leas in front of the building peering into the glasshouse to catch a glimpse of him. Because he and his friends were heavily bearded, it became likened to looking at monkeys in a cage, hence it became known as the Monkey House.

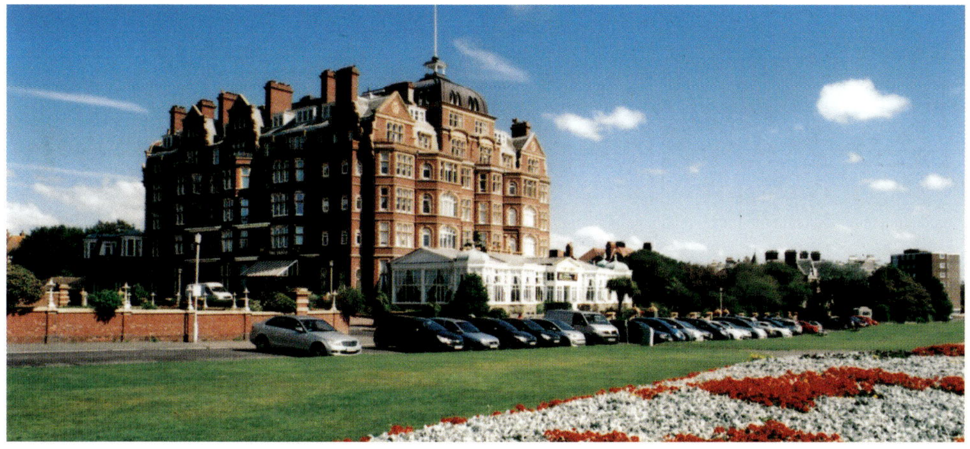

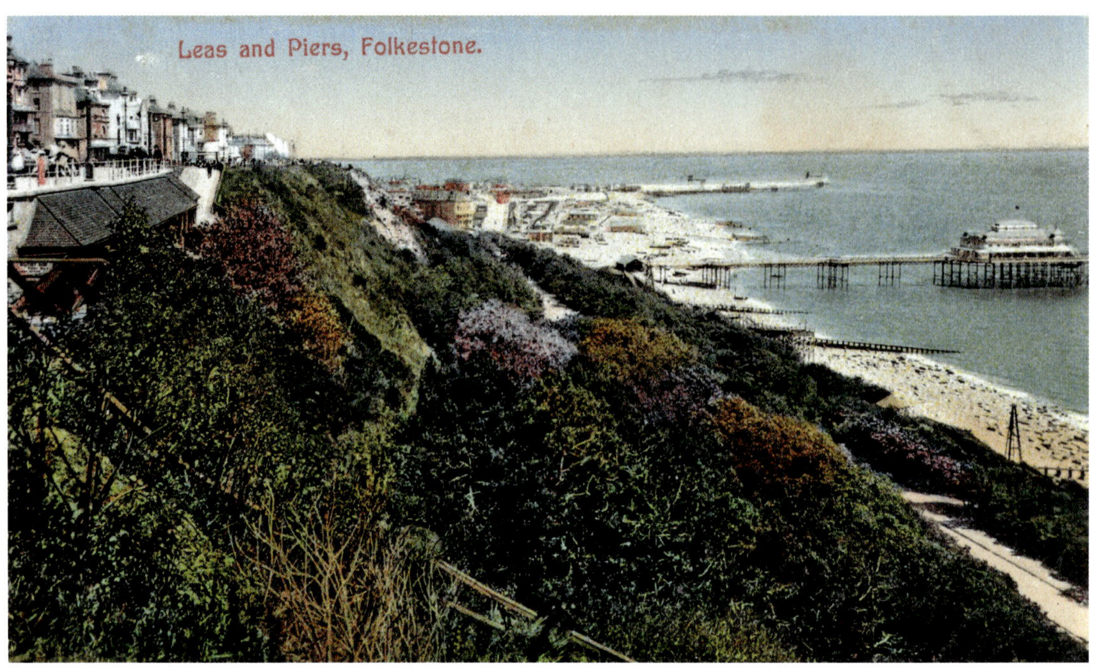

Folkestone's Coastline

Looking back along Folkestone's coastline towards the harbour and the pier early in the twentieth century. Also to be seen are some of the many footpaths which snake down the cliff from The Leas to the beach. The major difference in the two views is the loss of the pier.

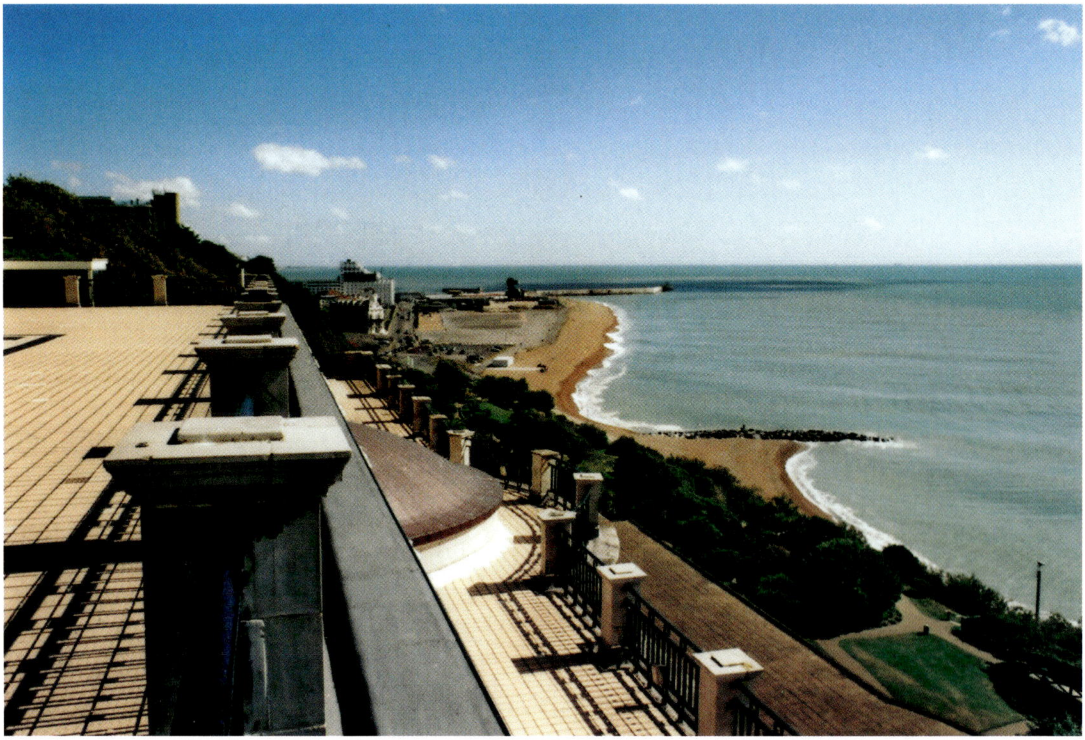

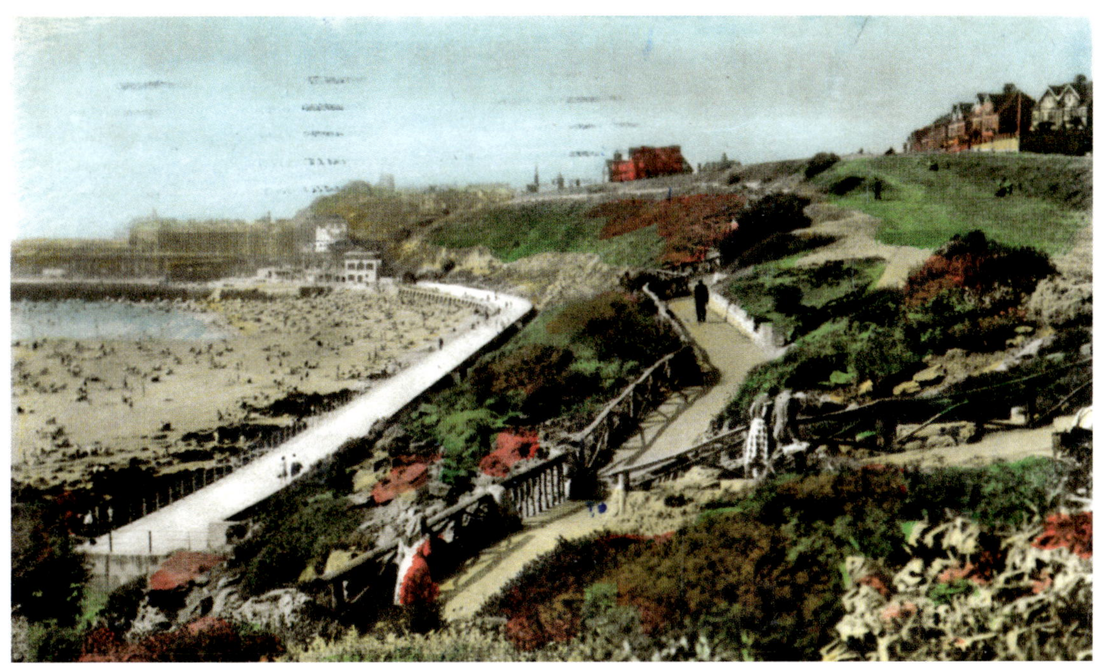

East Cliff and its Rock Garden

These are the rock gardens up on the East Cliff, but unfortunately they are no longer there due to cliff erosion and sea defence works. The early view shows they were still an attractive seafront feature until as late as 1954.

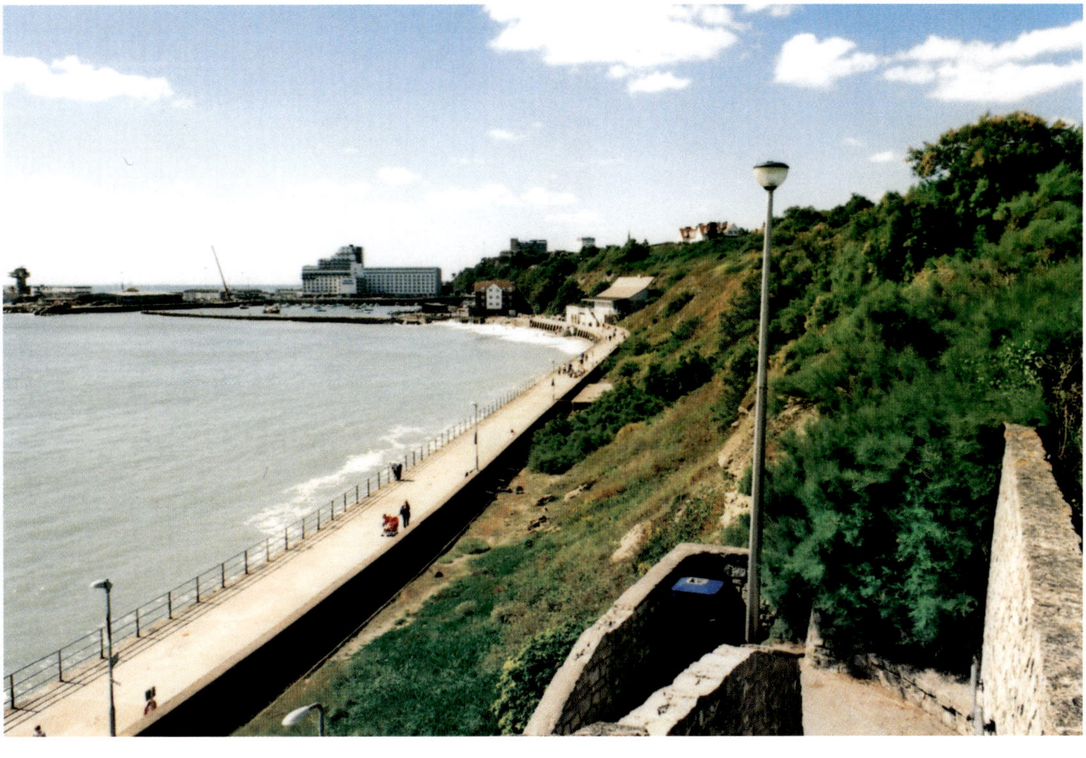

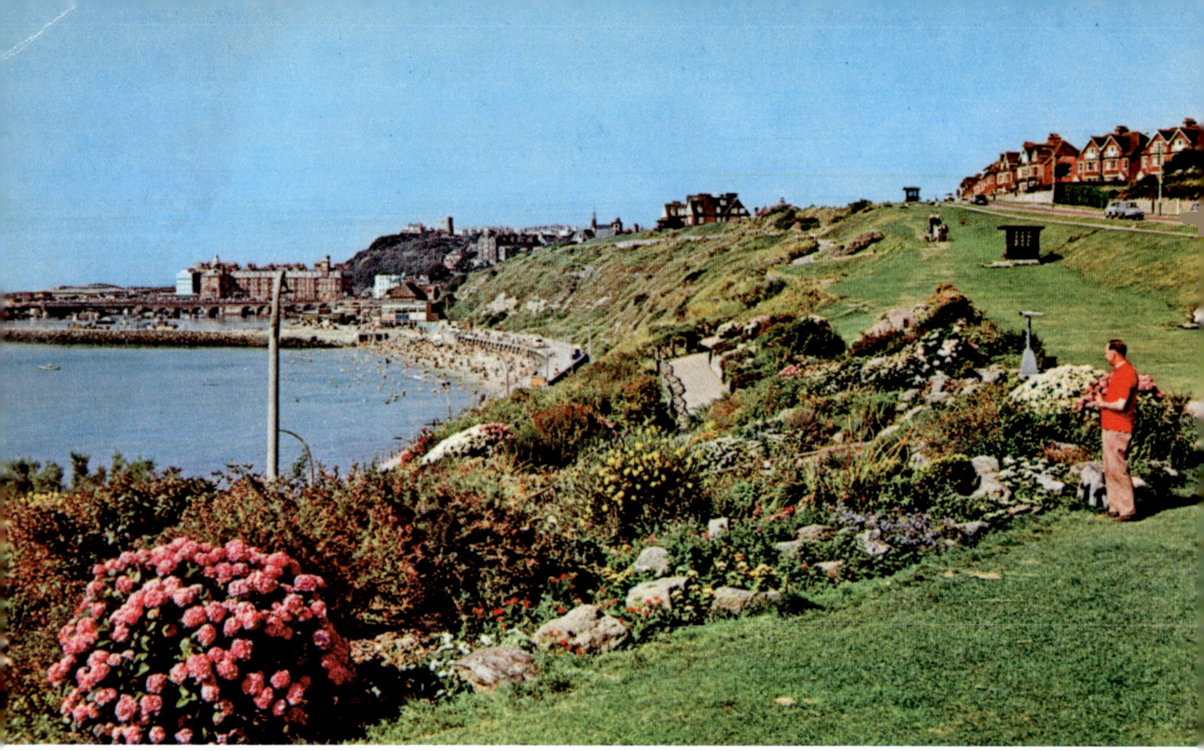

East Cliff and its Rock Garden
Another view of the East Cliff rock garden, only this time taken from further up the cliff.

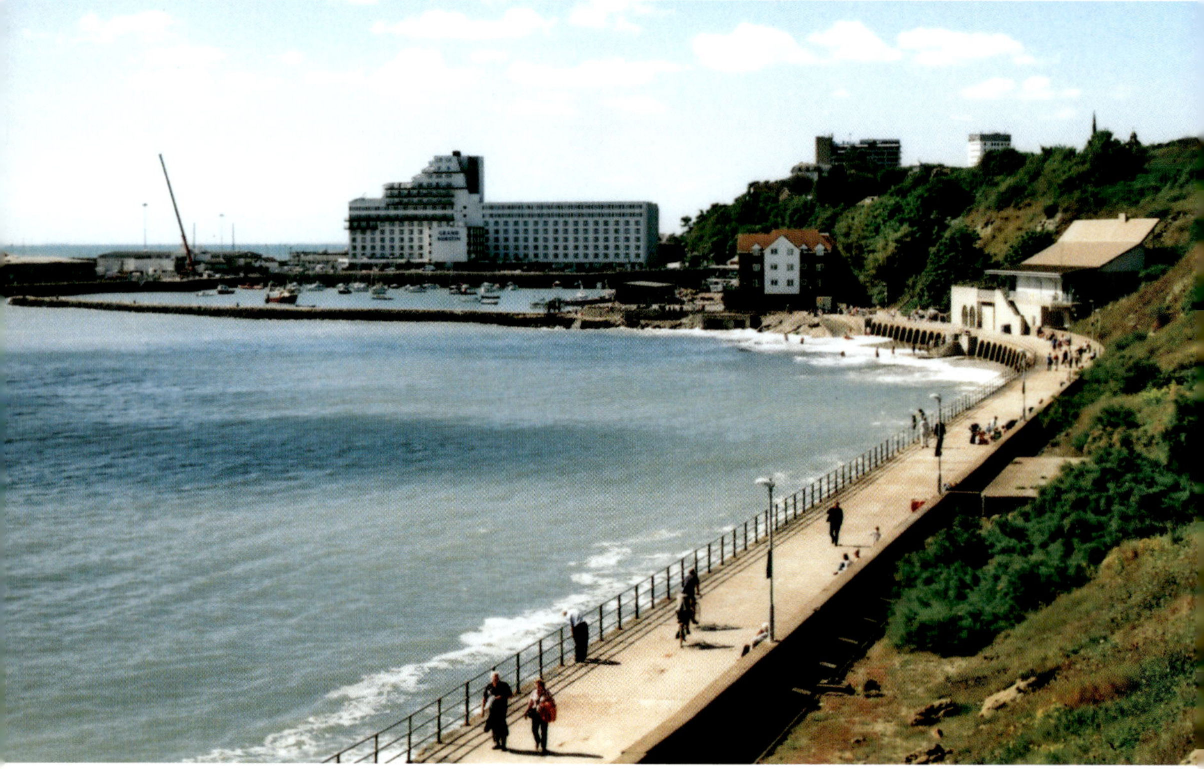

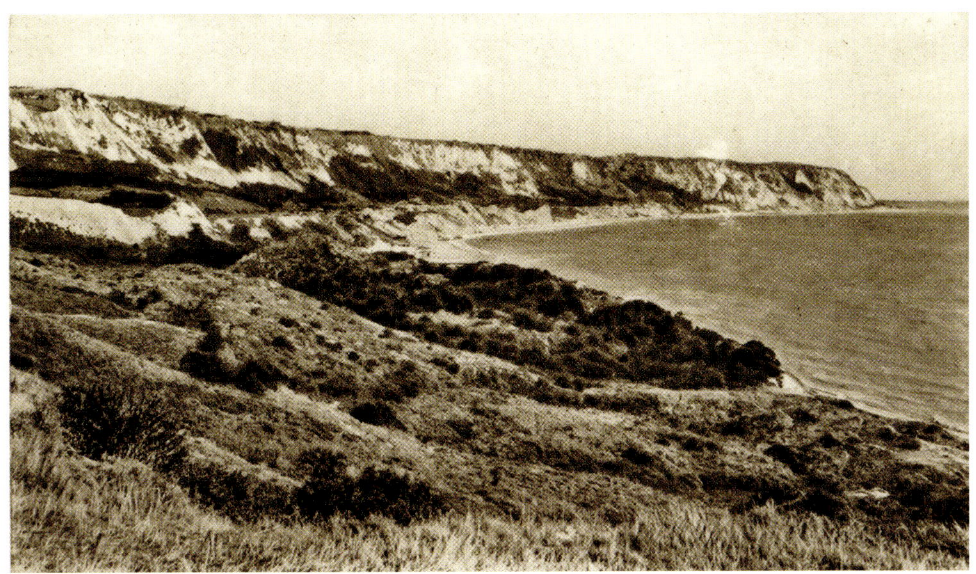

The Warren

The section of coastline to the east of Folkestone is known as the Warren and is a Site of Special Scientific Interest. The Warren is made up of a number of extremely ancient rocks, including chalk and Gault clay that has caused a series of landslides over the centuries. For this reason the area has probably been more intensively studied than any other landslide area of comparable size in Great Britain. This is largely because it is crossed by the main Folkestone–Dover railway line, which on occasion has been displaced by slipping, notably in 1915, creating an immediate demand for detailed studies and monitoring. The site has suffered twelve major slips since 1765, and is now protected by a complex of coastal defence works.

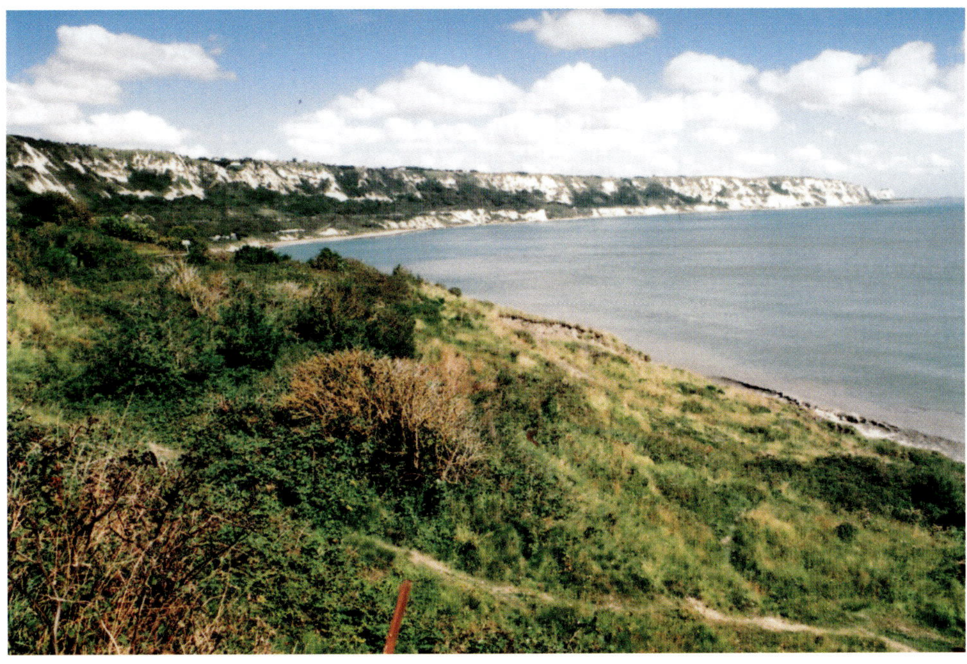

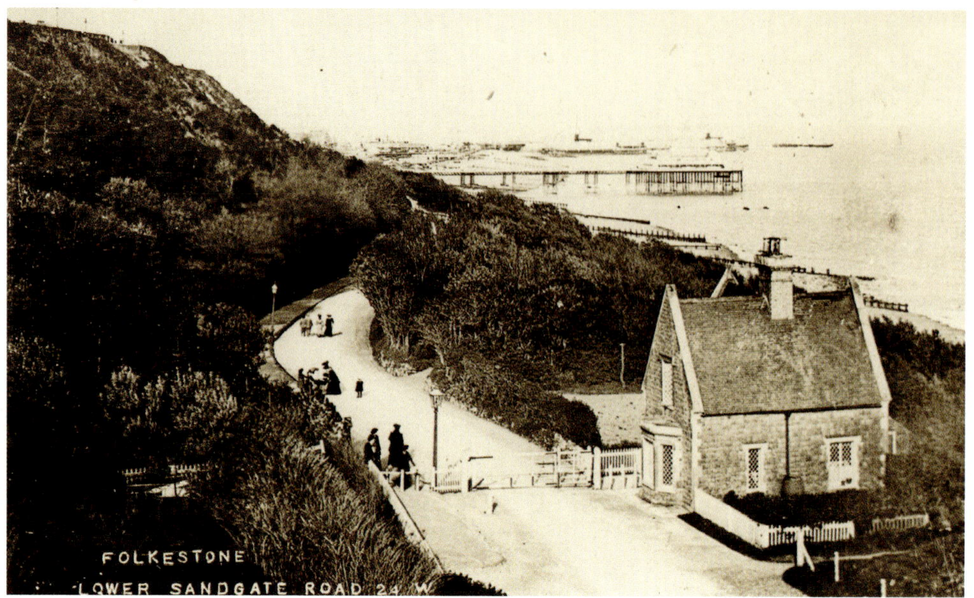

The Sandgate Road Toll-Bar

To the west of Folkestone is the suburb of Sandgate. To get there, all traffic had to negotiate a steep and winding hill until there was a major landslide at an unknown date, creating the area now known as the Lower Leas. A road was built along the seafront, creating easier access to Folkestone harbour, and there became the need for a toll-house. Originally this was a privately owned toll-road and payment would be collected from all vehicles travelling along that route. Collecting the levies kept someone gainfully employed and kept the road in good repair for many years. The toll-house is probably the most photographed house in Folkestone, even though it is now a private residence. Today the Lower Sandgate Road has gone, having been turned into the Lower Leas Coastal Park, which includes a playground for children, paths, flowerbeds and shrubberies. It is a beautiful area in which to spend a summer afternoon.

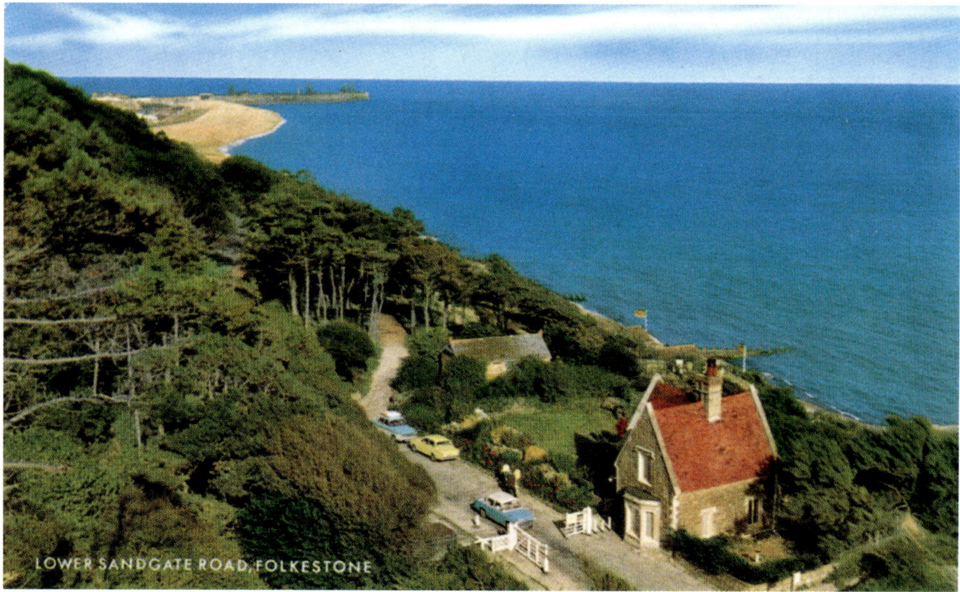

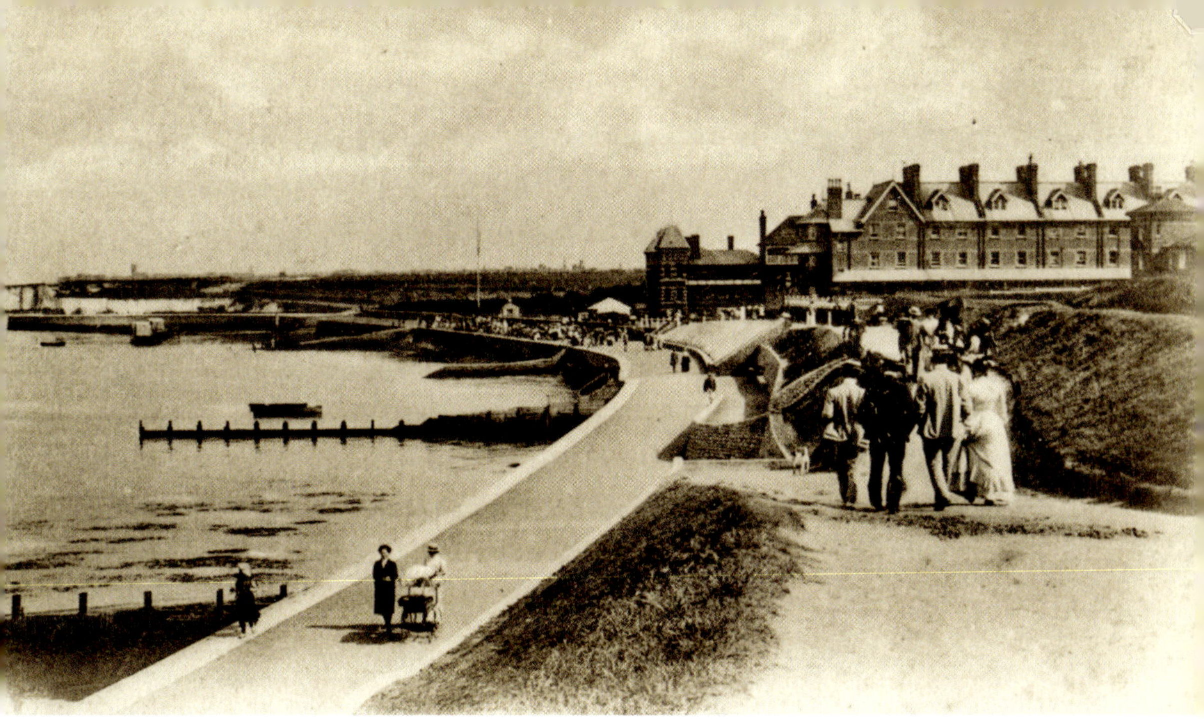

The Little-Known Bays and Beaches

Scattered all around the Kent coast are a number of delightful little semi-secluded bays and beaches, like St Mildred's Bay at Westgate-on-Sea (above). It has changed little in the century and more that separates the two pictures (below).

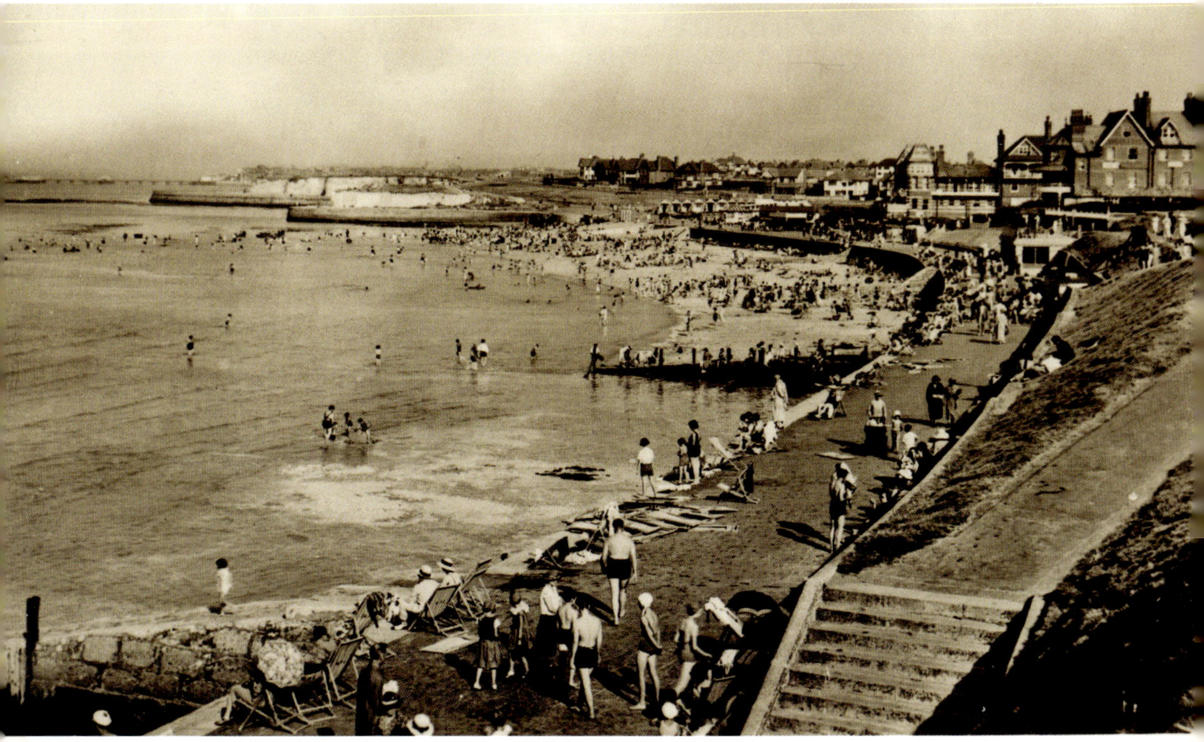

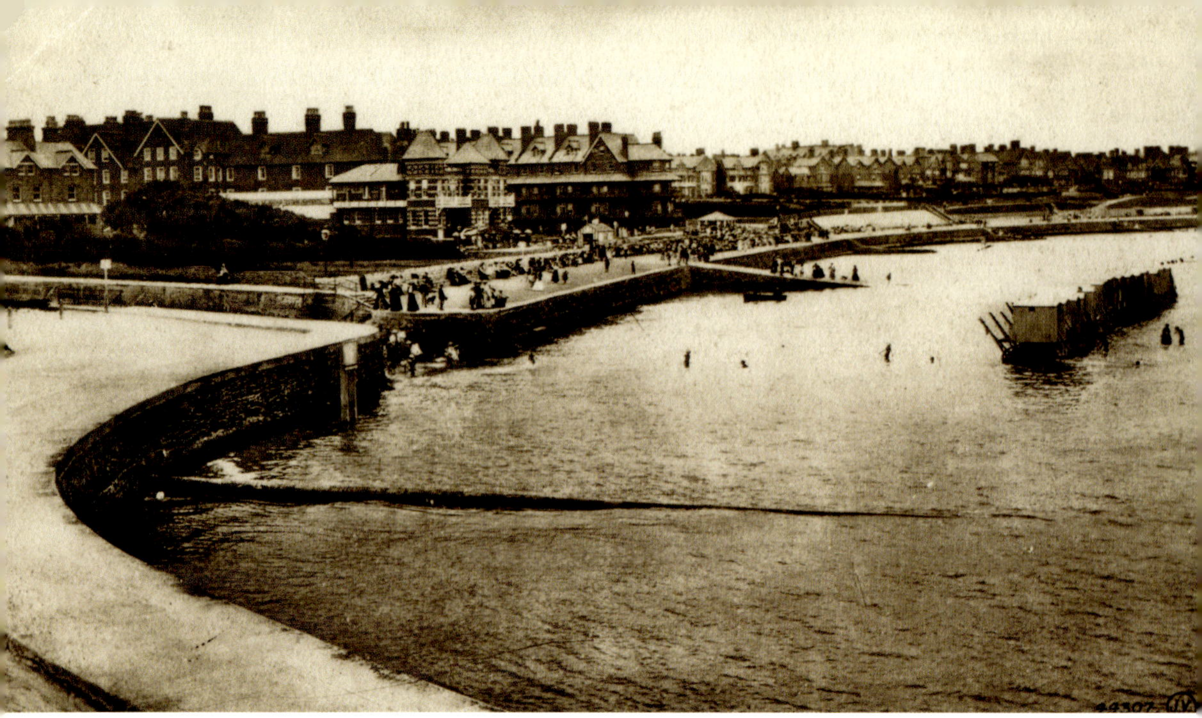

St Mildred's Bay

Such was the popularity of St Mildred's Bay that it even had its own bathing machines, something only normally found in the larger resorts. Below is another bay popular with day-trippers, Minnis Bay at Birchington, seen in this 1960s view of its promenade.

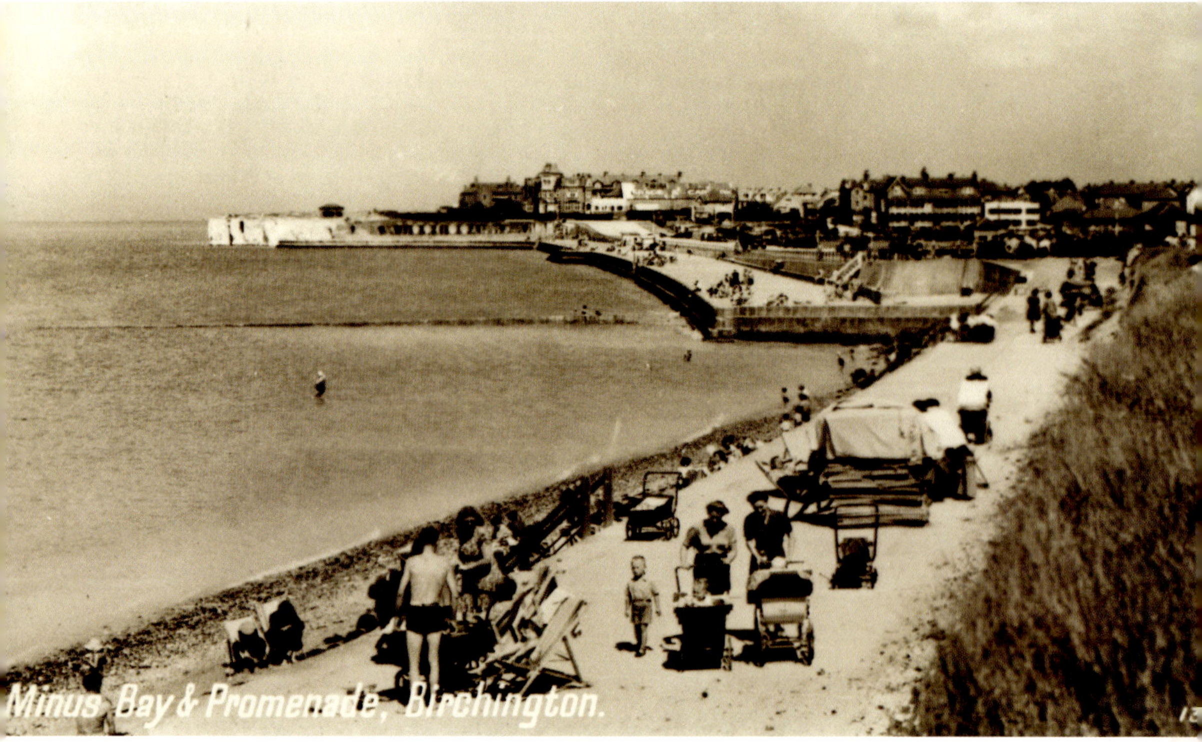

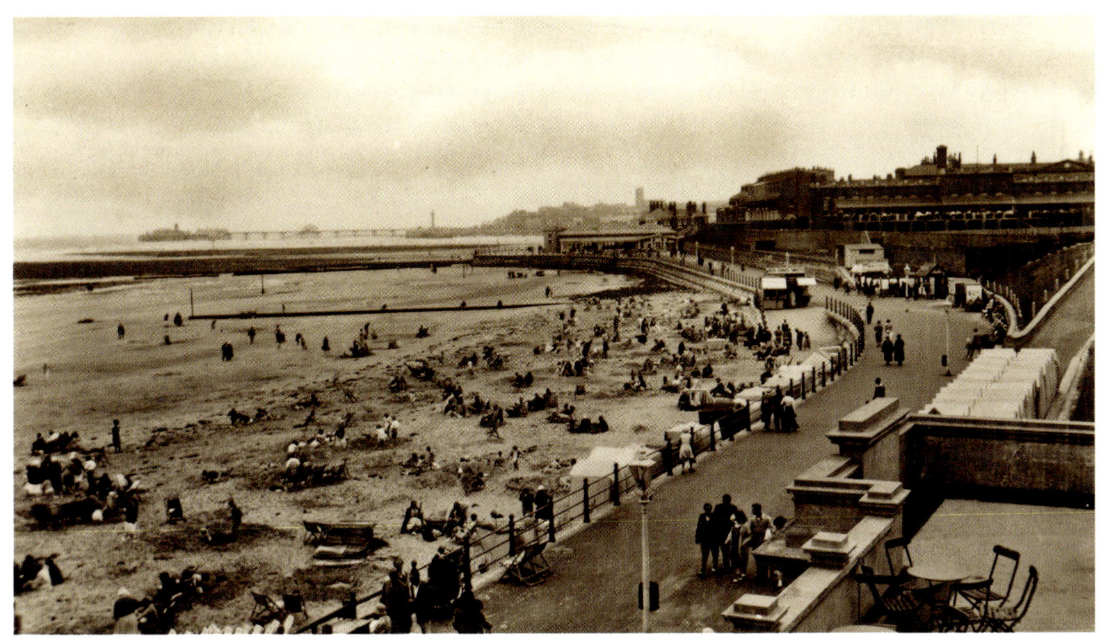

Margate

I well remember spending a few family holidays in Westbrook to the west of Margate. In this view Margate Pier can clearly be seen stretching out to sea, as can the original lighthouse, which was much taller than today's replacement. To the right of this 1920s/30s view can be seen the Royal Seabathing Hospital. Palm Bay, near Cliftonville, is something of a sun-trap, nestling beneath the chalk cliffs. It was without doubt extremely popular with the Edwardian day-trippers who, according to this picture, flocked there. The green fields above have now been lost to housing development.

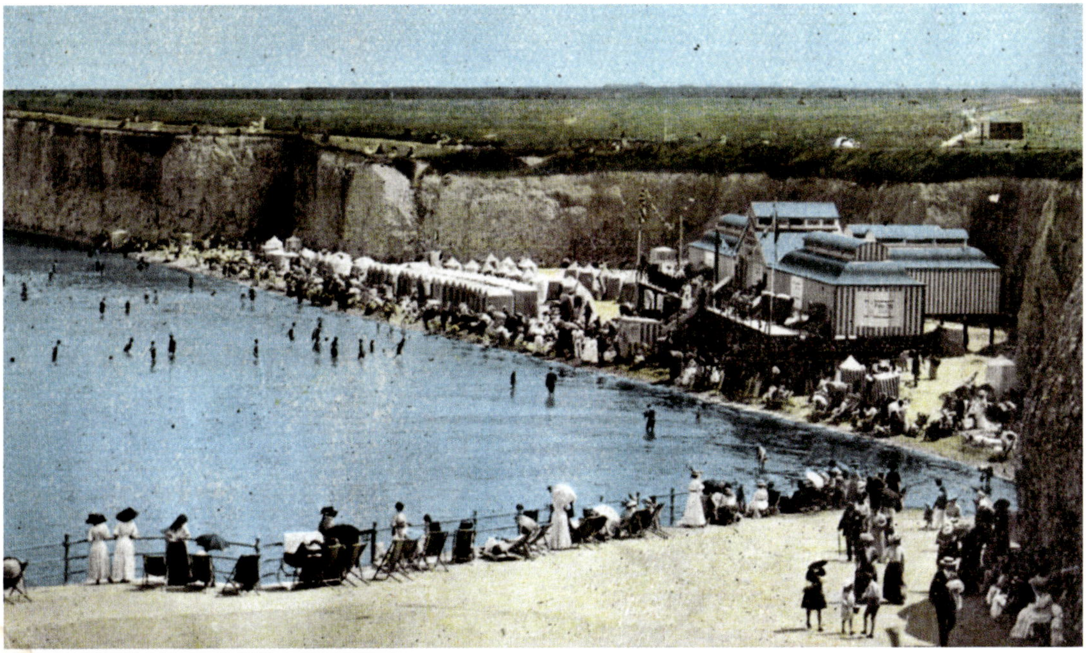

Acknowledgements

I would like to record my sincerest thanks and gratitude to the following people for their help and assistance in compiling this book. Without their help it would not have been possible to present a full and accurate picture of the changes that have taken place to that seaside icon, the pier and promenade, over the years. Firstly to Campbell McCutcheon who, having heard of my proposal for this book, immediately offered some old postcards to get my collection started. Thanks also to Charlotte Harlow; Suzannah Foad and the Margate Civic Society; Barry Kinnersley; Mrs Pat Clancy; Mrs J. Halligan.

Acknowledgement should also be made of the work of the early postcard photographers, many of whose names have long been forgotten, but without whom we would not have such splendid early views. In some cases many of the scenes they captured have not changed much in a century or more; only close inspection reveals the subtle changes that have taken place. These old postcards serve to remind us of how things once were.

Mention should also be made of those who sent the postcards; often their cryptic message has a story to tell. The fashion for sending picture postcards to friends and family a century ago and more has left us with the rich legacy of a window peering into everyday life as it was then. They can be compared to today's e-mail and mobile phone text messages, only longer lasting. Every part of a postcard from the picture to the stamp, the postmark, the message it contains and the address is a reminder of a time gone by and a valuable record of our social history. Postcards have been described as being 'the small change left over from art and poetry, but sometimes this small change suggests the idea of gold'.

Every effort has been made to authenticate the information contained herein, but sometimes time can play tricks on the memory so if any part of my text is inaccurate I sincerely apologise and will endeavour to correct it in any future publications of this book. Every reasonable effort has been made to contact the copyright holders of the photographs and illustrations used herein but should there be any errors or omissions or people whom I could not contact, I will be pleased to insert the appropriate acknowledgement in any future publication of this book.

John Clancy BA (Hons), MA.